The Use of Stereographic Projection in Structural Geology

Third edition

F. C. PHILLIPS, M.A.

Emeritus Professor of Mineralogy & Petrology
University of Bristol

EDWARD ARNOLD   LONDON
CRANE, RUSSAK   NEW YORK

© F. C. Phillips 1971

First published 1972 by
Edward Arnold (Publishers) Limited, London

Distributed in the United States of America by
Crane, Russak & Company Inc.,
52 Vanderbilt Avenue,
New York, New York 10017

Library of Congress Catalogue Card Number 72-76952

ISBN 0 8448 0005 8

All Rights Reserved. No part of this publication may be reproduced, stored in a retrieval system, or transmitted, in any form or by any means, electronic, mechanical, photocopying, recording or otherwise, without the prior permission of Edward Arnold (Publishers) Ltd.

Printed in Great Britain

PREFACE

MANY of the problems which arise during a study in structural geology are essentially exercises in three-dimensional geometry. Data acquired from field-observations—strike and dip of planes (bedding-planes, planes of schistosity or cleavage, faults, etc.), the pitch and plunge of lineations (outcrops, slickensides, micro-folds, flow-lines, elongation of mineral components or of inclusions, etc.) and all other related vector data are incorporated in a geological map and in accompanying illustrative sections. Both in the construction of such sections and of related block-diagrams, and in the further study of the structural features of an area, it is frequently necessary to make various calculations. Sometimes, indeed, as in many of the problems of mining geology and in the interpretation of borehole cores, the necessity for calculation arises at an even earlier stage, since such fundamentals as the dip and strike of a bedded series are only indirectly derived from the observed data.

Methods of calculation most generally used in the past have involved the application either of descriptive geometry or of trigonometry. Many ingenious devices in the form of tabular or graphical aids to such solutions have been devised, but an outstanding drawback remains that the necessary 'three-dimensional' figures are in many instances complex and difficult to construct. A geologist who has taken a course of training in crystallography and mineralogy turns naturally to stereographic projection as affording the neatest representation of three-dimensional geometry and combining ease of solution with extreme speed. In the Introduction to the first edition of this book, published in 1954, the comment was made 'With expansion of the whole field of geological inquiry, however, it has come about that many geologists who eventually become involved in structural studies have little or no acquaintance with the stereographic method.' This position has changed radically in the past fifteen years. It must be rare indeed to-day to encounter a qualified geologist who has not had a thorough grounding in structural geology, and equally rare must be a syllabus which does not include at least general reference to stereographic procedures.

Scattered through older geological literature are a few special papers designed to draw the attention of the general geologist to the value of the stereographic technique. A paper by Bucher (1920)*

* Dated references throughout are to the Bibliography, pp. 79–83.

on a particular application was followed twenty-four years later by a more general account (1944). The method was also in use at this time on the Continent (e.g., Seitz, 1924; Wegmann, 1929, 1929b) and a useful book by Sokol (1927), apparently little known in this country, makes many references to its application. Within a rather specialized field, the school of structural petrologists developed by the work of Sander (1930, 1948, 1950) has from the outset made extensive use of this and of closely allied procedures. In America, Fisher (1938, 1941, 1941a) has been an ardent proponent, and other more recent individual contributions on particular applications are listed in the Bibliography. Billings (1942) made brief reference to the possibilities, but the first general account in a standard American text-book appeared in the fourth edition (1949) of Nevin's valuable *Principles of Structural Geology*. In the changed situation noted above it has become standard practice to incorporate an account of stereographic techniques, offering considerable detail, in text-books of structural geology (see, for example, Donn and Shimer, 1958; Badgley, 1959; Ragan, 1968).

I have made constant use of stereographic procedures in my own research for many years, and preliminary drafts of the present work were used with classes of Honours Students in the University of Bristol from 1948 onwards. In 1952, Mr. A. W. Kleeman of the University of Adelaide drew my attention to the fact that Prof. L. A. Cotton made considerable use of the method as part of a course in geological mapping at theUniversity of Sydney, and that he and Dr. Garretty had compiled a manuscript. Though never published, this was issued in Roneoed form as Bulletin No. 2 (1945) of North Broken Hill Ltd., for private circulation, and has become fairly widely known in Australia. I am greatly indebted to the authors for sending me a copy of this Bulletin, and gratefully record its value to me whilst drawing up a definitive version of my own manuscript for publication; several of the exercises at the end of this book are based directly on examples given by these authors. I have also received helpful advice and criticism from Dr. Gilbert Wilson, Reader in Structural Geology in the University of London.

In the second edition of this present work, published in 1960, no major changes were made. This third edition is now offered more as an introductory text-book at undergraduate level than as a complete piece of propaganda for its subject (an objective which has so largely been attained). For this very reason I have avoided overloading it with much new material and added complexities, though reference is made to important developments such as the continuous dipmeter. The Bibliography is enlarged to include recent papers which elaborate the scope of some of the applications beyond the limits of this book. Now that almost any published tectonic study is likely to illustrate analysis and synthesis of its data by using stereographic or equal-area projections, it has become unnecessary to include many examples. A few relevant text-books are cited. As with earlier editions, it remains my earnest hope that the book will continue to be found an acceptable and useful addition to geological literature.

Brockenhurst, 1971   F. C. PHILLIPS

CONTENTS

	PREFACE	iii
I	THE PRINCIPLE OF STEREOGRAPHIC PROJECTION	1
II	TRUE AND APPARENT DIP	7
III	INTERSECTING INCLINED PLANES	12
IV	ROTATION OF THE SPHERE	28
V	ROTATION ABOUT INCLINED AXES	36
VI	STEREOGRAPHIC PROJECTION AS AN AUXILIARY TO FINAL SOLUTION	54
VII	TECTONIC SYNTHESES IN STEREOGRAPHIC (AND RELATED) PROJECTION	60
	APPENDIX—CALCULATIONS BY SPHERICAL TRIGONOMETRY	70
	BIBLIOGRAPHY	79
	EXERCISES	84
	ANSWERS TO EXERCISES	86
	WULFF NET	87
	INDEX	89

CHAPTER I

THE PRINCIPLE OF STEREOGRAPHIC PROJECTION

SUPPOSE that at a point *O* on the surface of the earth (Fig. 1) a structural plane is observed striking in a direction 300° and dipping 50° south-westerly. (It is now a fairly firmly established convention in structural geology to quote geographical directions in terms of the 360° compass-rose, setting the 0° due north and graduating clockwise. If the bearing (azimuth) of the direction of *dip* is stated, e.g., 210° in this example, the simple statement is unambiguous; if,

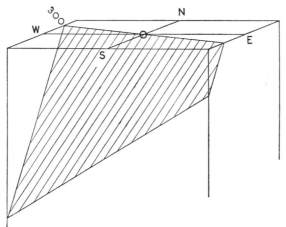

FIG. 1   Block-diagram of a plane, strike 300° and dip 50°.

as is sometimes convenient, the direction of strike is stated we must add to our statement of the amount of dip a phrase such as 'south-westerly' to distinguish between two possible tilts in opposite directions.) If a sphere be inscribed about *O* as centre, the structural plane will intersect the surface of this sphere along a great circle (Fig. 2). A *great circle* is thus defined, as the intersection of a sphere by any plane passing through the centre of the sphere. The problem of representing this *spherical projection* of the plane on a two-dimensional projection is similar to that of the astronomer in producing a map of the celestial sphere or of the cartographer in preparing a map of the world.

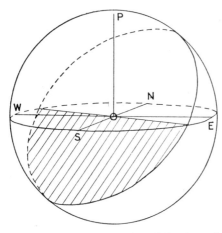

FIG. 2   Spherical projection of the plane of Fig. 2.

The particular device of *stereographic projection* is a very ancient one, already in use in Greece in the second century B.C.; it achieved great popularity in the hands of crystallographers, who developed it highly for the study of crystal morphology and crystal optics. Such a projection, on an equatorial plane of the sphere, of the lower half of the great circle of Fig. 2 is obtained by joining all points of the semicircle to the zenithal point P of the sphere and marking the intersections of these joins with the equatorial plane (Fig. 3). The resultant projection, Fig. 4, is an example of a simple *stereogram*. (This word, universally used by crystallographers with this meaning, is widely applied both on the Continent and in America, to block-diagrams; it also appears in the literature of aerial photography as the name for a 'stereo-pair' of photographs correctly mounted for stereoscopic observation.)

The limiting circle, the circumference of the projection, is called in crystallography the *primitive*. It will be clear from Fig. 2 that if points on the upper half of the sphere were to be projected from the zenithal point P in like manner, their projections on the plane of the stereogram would lie outside the primitive; in particular, the projection of a point on the surface of the sphere close to the point P would lie at an inconveniently great distance beyond the primitive. We shall, however, find it unnecessary at present to project more than the lower hemisphere, and can postpone further consider-

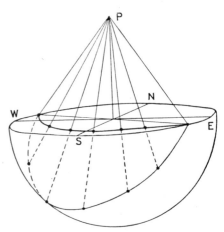

FIG. 3   Stereographic projection of the plane of Figs. 1 and 2.

THE PRINCIPLE 3

ation of this practical difficulty. (If the reader should be already acquainted with the customary stereographic procedure of crystallographers he will be aware that they usually project only the *upper* hemisphere; we shall have to consider this point again later, by

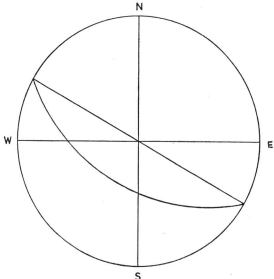

FIG. 4 The completed stereogram of the plane.

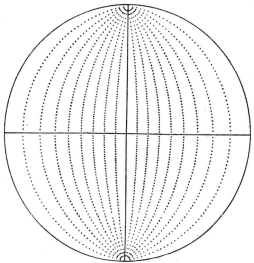

FIG. 5 Stereogram of a family of meridional great circles.

which time it will be clear that in most geological problems it is preferable to adhere to projection of the lower hemisphere, a convention now firmly established also in the specialized field of structural petrology.)

One of the many valuable properties of stereographic projection lies in the fact that when a great circle is projected as we have described the resultant curve in projection (Fig. 4) is still an arc of a circle. By projecting in this manner a series of planes striking N–S

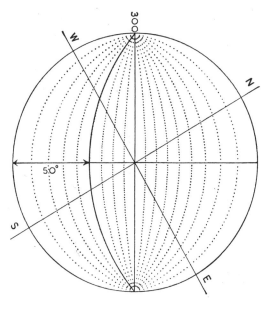

FIG. 6 Tracing the plane with strike 300° and dip 50° south-westerly.

and dipping E (or W) at various angles we can construct a 'net' of meridional (great circular) curves as in Fig. 5, in which the planes are drawn for every 10° of dip between vertical and horizontal. If now a sheet of tracing paper is placed above such a net we can trace the projection of a plane with any stated direction and amount of dip by revolving the tracing paper above the net to the appropriate orientation. Fig. 6 is drawn to represent a net with the common strike line of the meridional planes oriented N–S on the paper; on the tracing paper above it reference lines NS and WE have been drawn, and the tracing paper has been revolved until the 300° strike line lies in such a position that the plane of Figs. 1–4, dipping south-westerly at 50°, can be traced. (We shall use the term *revolve* for this action whereby the tracing paper is turned above the net about its centre; the term *rotation* will be introduced later, when it becomes necessary to re-orientate the spherical projection in space by turning the sphere about some specified diameter, before constructing the stereogram.)

Not only great circles, but small circles also, project stereographically from the surface of the sphere as circular arcs in projection. (A *small circle* is the intersection of a sphere by any plane not passing through the centre of the sphere.) Suppose that with centres N and S on our original spherical projection we inscribe a series of small circles of increasing radius (a few are shown diagrammatically on the lower hemisphere in Fig. 7). Stereographic projection of these yields a number of small circular arcs on our net (Fig. 8) which graduate the meridional great circles by their intersections—we have evolved the standard *meridional stereographic net* (often called by crystallographers a 'Wulff net', after G. V. Wulff who published a reproduction of such a net in 1902). By tracing from a net we shall carry out the great majority of useful constructions without further graphical aids.

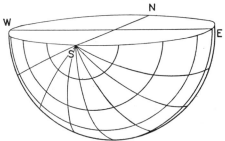

FIG. 7   Small circles, centred at S, drawn on the lower hemisphere.

The simplest arrangement in practice in the laboratory is to work with a sheet of tracing paper revolving above a fixed net. A printed net is pasted to a suitable foundation such as thick card, thin plywood, plastic or metal, and a large drawing pin is passed through the centre of the net from below. The tracing paper revolves about the projecting pin, its central area being strengthened if necessary by a small patch of surgical tape or *Sellotape*. Nets 20 cm. diameter, with great and small circles at 2° intervals printed on white paper or card, can be obtained from Vickers Instruments, Haxby Road, York YO3 7SD, or from the *Geological Journal* (The Treasurer, Geological Journal, Department of Geology, The University, Liverpool). The former also supply a permanently mounted model, in which the net revolves beneath a fixed frame to which the tracing paper is fastened; thus the N–S line on the projection remains stationary in its proper orientation throughout the work, a point which the beginner may appreciate, but most of our figures below are drawn to correspond with the simple arrangement of a fixed net. For use in the field a smaller net is more convenient, and the printed net at the end of this book may be cut out and mounted for the purpose, or nets 10 cm. diameter can be obtained from the *Geological Journal*. If the readings in the field are magnetic, the necessary correction can be made once for all when plotting by an appropriate offset of the north index mark on the tracing paper.

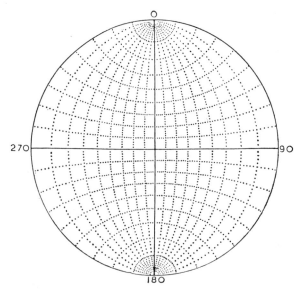

FIG. 8   A 'Wulff' stereographic net.

CHAPTER II

TRUE AND APPARENT DIP

REFERENCES:—Sokol, 1927, p. 151; Hubbert, 1931; Earle, 1934, Chaps. I, IX; Nevin, 1949, pp. 381–2; Stockwell, 1950; Wollnough and Benson, 1957; Donn and Shimer, 1958, pp. 132–6.

IN any vertical section not normal to the strike of a plane, the dip of the trace of the plane (*apparent dip*) is less than the true dip. The diameters NS and WE of the stereographic net (the *principal diameters*) represent, as circles of infinite radius, the traces of two vertical great circular sections of the sphere (Fig. 9). The plane represented in Fig. 10 dips at an angle of 30° on a bearing of 210° (strike 300°); it is projected stereographically in Fig. 11. The apparent dip to the west is read from the arc *ab* as 16°; the apparent dip to the south (*cd*) is 26½°. To determine the apparent dip in any other direction, draw a radius of the projection in the appropriate azimuth and read off the required arc by revolving the tracing above the net until the arc lies above a principal diameter. The apparent dip of the plane in Fig. 11 to the south-west (direction 225°) is thus determined as 29°. Such determinations of apparent dip are necessary, for example, when constructing true-scale 'horizontal sections' of a geological structure when the plane of section is not normal to the strike (see p. 56).

The converse problem of determining the direction and amount of true dip from measurements of apparent dips is also readily solved. Points *b* and *d* (Fig.12) are plotted to show two apparent

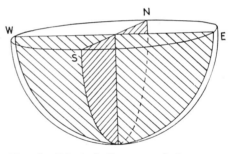

FIG. 9   Principal diameters of the net as projections of vertical planes.

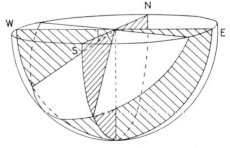

FIG. 10   Apparent dips of an inclined plane to S and to W.

dips of a structural plane correctly in azimuth and amount (b 13° at 295°, d 20° at 200°). The tracing is now revolved above the net until b and d lie above a common great circle (drawn full in Fig. 12), when the true dip is at once determined in amount and direction from the arc fg (24° on a bearing of 236° in our example). In the simple text-book version of this problem the two apparent dips are often supposed to have been measured on adjacent quarry-faces; this perhaps savours rather of the artificial for it would usually be

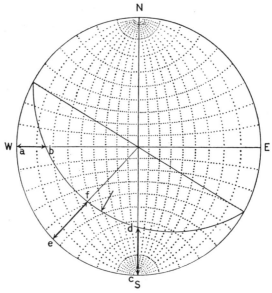

FIG. 11   Stereogram for determination of apparent dips.

possible in a quarry to lay bare an actual dip-surface. Apparent dips, however, may well be all that can be observed when sighting outcrops on cliff faces, canyon walls or distant hill slopes, but here a further point arises—our line of sight is unlikely to be accurately normal to the face exposing the outcrop, so that the angle which we observe is only a projected value of the actual angle of apparent dip on the face of the exposure. Even where the exposure is directly approachable the measurable traces of the bed may not lie on vertical planes (as, for example, where a bedding plane intersects two directions of non-vertical cleavage). We shall return to these more generalized problems when we have further developed our technique.

The 'three-point problem' is another elementary text-book exercise in the determination of true dip from apparent dips. The height of a structural plane referred to datum level is known at three mapped points, either from observations of surface outcrops or from borehole logs. Alternatively, the data may be derived from a dipmeter survey of a single borehole (p. 51). From these data, the slopes in two known directions are determined as gradients, and by converting the gradients to angular values of apparent dip we can proceed as before. The conversion may be done graphically or from a table of cotangents; Earle (1934, p. 13) draws attention to a useful approximation for small angles.

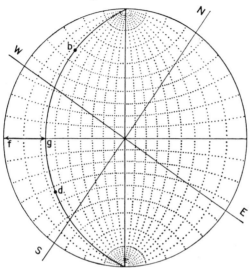

FIG. 12   Stereographic determination of true dip from two apparent dips.

Lineation on planes

REFERENCES:—Ingerson, 1942; Fisher, 1943; Bucher, 1944; Sander, 1948; Clark and McIntyre, 1951; Billings, 1954.

When examining in the field a particular structural plane of ascertained strike and dip, we may frequently observe upon it some kind of *lineation*. On a fault-plane there may be a direction of slickensides, on sedimentary bedding-planes the traces of intersections of joint- or cleavage-planes, in igneous rocks the traces of a flow-structure, and on metamorphic rocks any one or more of the many types of lineation studied in detail by structural petrologists. Working towards a fuller understanding of the geological structure we must measure and record all such lineations, and here it becomes necessary to distinguish carefully between the related concepts of *pitch* and *plunge*.

In Fig. 13 *OSQR* represents a lineated structural plane dipping due east. We may record the attitude in space of the lineation by measuring the angle *QOS* which it makes with the recorded strike of the plane (the angle β in the figure*) which is termed the *angle of*

*Sander, 1948, p. 136, uses the symbol ζ for the angle of pitch.

pitch (or *rake*) of the lineation within the plane. Alternatively, we may measure the angle *QOP* (the angle α) in a vertical plane—the *angle of plunge* of the lineation—and record also the bearing of the line *OP*—the *trend* of the lineation. It is inadvisable for several reasons to restrict attention to either one or other of these related concepts. Whilst it is often possible to measure the trend and plunge directly in the field, and special types of clinometer-compass have been designed to facilitate this operation, on some kinds of outcrop (and especially on steeply-dipping planes) a more accurate result is obtained by

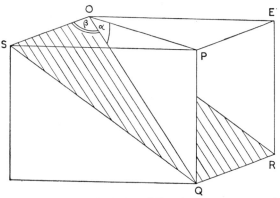

FIG. 13   Pitch and plunge of lineation on an eastward-dipping plane.

indirect determination of the plunge from measurements of pitch on planes of known strike and dip (Ingerson, 1942). Moreover, though the trend and plunge are fundamental and will in general be shown on a structural map, in some problems it is the pitch that is of more immediate practical interest; a mining geologist, for example, follows an ore-shoot down a dipping vein exclusively in terms of the angle of pitch.

In Fig. 14 the plane *OSQR* of Fig. 13 is plotted stereographically, dipping due east at 17°. The arc *SQ*, read off along the meridional great circle by means of the graduating small circles, represents the measured angle of pitch within the plane ($\beta = 54°$). Joining the point *Q* to the centre of the stereogram, the trend is given by the position of the point *P* (127°, or E 37° S) and the angle of plunge by the arc *PQ*, this value being read by revolving the tracing until *PQ* lies along a principal diameter ($\alpha = 13\frac{1}{2}°$). (Since a stereogram is a representation of the *surface* of a sphere it is not formally correct to represent a lineation as a series of lines, as in Fig. 14. The point *Q* alone should mark the emergence of the direction of lineation through the sphere. For purposes of demonstration, however, it sometimes pays to behave informally and to sketch into a stereogram a feature such as a lineation as it would be seen on looking at the plane itself.)

This simple example demonstrates the relationship between the trend and plunge of a lineation and its angle of pitch on a plane of known attitude. When mapping an area of well-lineated rocks it is worth while to carry in the field a small stereographic net furnished with an overlay of matted transparent plastic. Readings conveniently

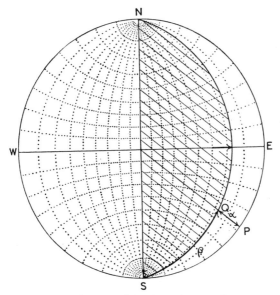

FIG. 14 Stereogram of the lineated plane of Fig. 13.

taken as angles of pitch can then be converted on the spot to trend and plunge and the appropriate symbol inserted at once on the field-slip. In this way, mistaken readings can usually be noticed at once as discordant and be revised; it is, however, wise to record the data of pitch in the field notebook, since further suggestions concerning the cause of the lineation (whether due to intersections of planes, a linear structure within the rock, etc.) and concerning its structural significance (whether marking a direction of movement or flow, the axial direction of associated folding, etc.) can often be derived from further plotting in the laboratory (p. 20).

CHAPTER III

INTERSECTING INCLINED PLANES

REFERENCES:—Bucher, 1920; Earle, 1934, pp. 15–21; Bucher, 1944; Ingerson, 1944; Lowe, 1946; Nevin, 1949, p. 382; Wallace, 1950; Clark and McIntyre, 1951a; Den Tex, 1953; Fisher, 1958.

THUS far we have been concerned with only one inclined structural plane and its intersections with vertical planes. In many problems, however, such as those involving outcrops on an inclined surface, intersections of beds with joints, veins or fault-planes, mutual intersections of veins, intersections of bedding-planes in folded regions etc., the lines of intersection of two or more inclined planes are involved.

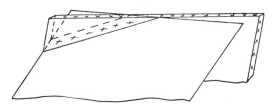

FIG. 15   A vein intersecting a dipping bedding-plane.

The trend and plunge of the line of intersection of two planes of known attitude are at once determinable from the point of intersection of their traces in stereographic projection. Thus in Fig. 15 a bedding-plane dipping 20° on an azimuth 200° is intersected by a vein dipping 60° on an azimuth 165°. Their stereographic traces in Fig. 16 intersect in the

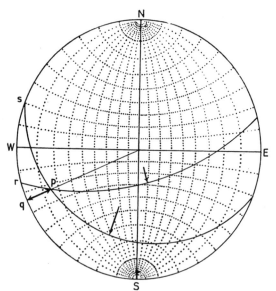

FIG. 16   Stereogram of the vein and bed of Fig. 15.

point *p*; drawing the radius through *p* to *q* we determine the trend of the intersection to be 247° (or W 23° S) and the plunge (from the arc *pq*) to be 14°.

In exactly similar manner, in a study of plunging ('pitching') folds

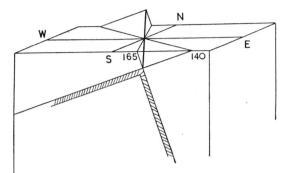

FIG. 17 Block-diagram of a plunging fold.

the attitude of the ideal 'axis' is determined from observations of dip and strike on either limb. In the asymmetric anticline of Fig. 17 the strike of the westerly limb is 140° and the dip 20° south-westerly, whilst the eastern limb strikes 165° and dips 70° easterly. From the stereogram, Fig. 18, the azimuth of the trend of the axis is 162° and the plunge 8° southerly.

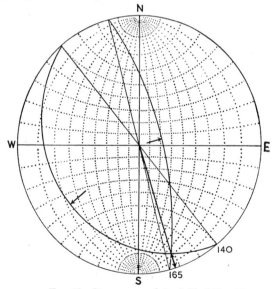

FIG. 18 Stereogram of the fold of Fig. 17.

We can now also extend our conception of apparent dip, since the apparent dip of one plane measured on another inclined plane is the same as the pitch of their line of intersection measured on that plane. In Fig. 16 the trace of the bed pitches 16° westerly (arc *pr*) on

STEREOGRAPHIC PROJECTION

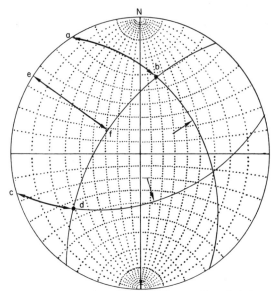

FIG. 19 Determination of true dip from angles of pitch on an inclined plane.

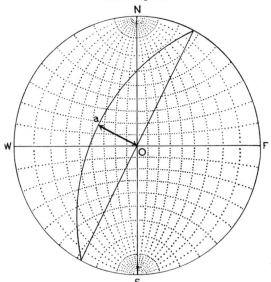

FIG. 20 Hade of a fault in stereographic projection.

the plane of the vein, whilst the trace of the vein pitches 45° westerly (arc *ps*) on the plane of the bed. It is no longer necessary for apparent dips, from which we obtain a true dip (p. 7) to be measured on vertical planes. We can use instead the pitch of traces on inclined planes. In Fig. 19 two cleavage planes, dipping as shown by the arrows, bear on their surfaces the traces of a bed, pitching at angles *ab* and *cd* respectively. The great circle *bd*, corresponding to the bed, and the dip *ef* are determined in the usual manner by revolving the tracing to bring *b* and *d* on a common meridian. Such measurements find frequent application in the construction of accurate sections projected on other than vertical planes, such as the projection of a vein-system on a fault-plane or a section at right angles to the axes of a plunging fold-system. In problems of this kind involving fault-planes it is important to remember that geologists frequently quote the *hade* of a fault, which is the complement of the dip; hence the angular value of hade is plotted outwards from the centre of the net—in Fig. 20 the fault hades 40° (length of the arc *Oa*) on an azimuth 298°.

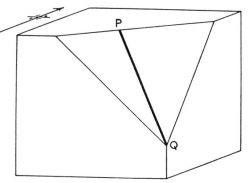

The use of intersecting inclined planes also enables us to return to the problem of plotting outcrops sighted from a distance (p. 8). In Fig. 21 the outcrop *PQ* runs across a surface

FIG. 21   An outcrop on an inclined surface.

sloping in a south-easterly direction. If we sight this outcrop from due south (Fig. 22) the actual angle of pitch of *PQ* on the face of the exposure is projected as the angle 1*mQ* on the vertical plane at right angles to the line of sight. This angle, however, also measures the dip to the east of the plane *PmQ*, striking N–S parallel to the line of sight and passing through the trace of the outcrop *PQ*. The value of 1*mQ* determined from a distant vantage-point is 58°. In Fig. 23 the vantage-point

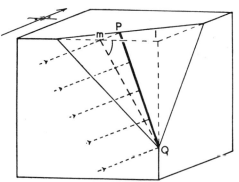

FIG. 22   Viewing the outcrop of Fig. 21 from the south.

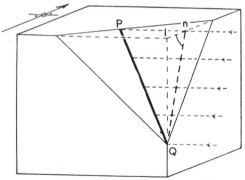

FIG. 23 Viewing the same outcrop from the east.

has been changed so that the same outcrop is now viewed directly from the east; the projected angle $1nQ$ is measured as 64°, which is thus the dip to the south of a plane striking W–E and likewise passing through the trace of the outcrop PQ.

Plotting these two planes stereographically (Fig. 24) the outcrop is determined as their line of intersection, trending E 38° S (128°) and plunging 52° south-easterly. If a second portion of outcrop of the same bed can likewise be sighted from two directions (the bed, for example, may show a V-outcrop across a valley, or be exposed both on the upper slopes and on the walls of a gorge) a second apparent dip can be accurately projected in similar manner and from these two the true dip is at once determined as on p. 7.

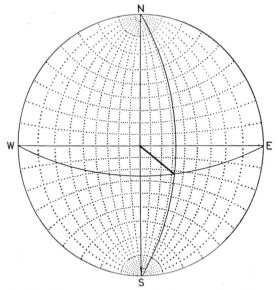

FIG. 24 Stereogram to determine the azimuth and plunge of the outcrop of Figs. 21–3.

The principle of this construction can be widely applied. In this example the lines of sight have been taken as running due N and W respectively for simplicity, but sighting directions of any measured azimuth can of course be used provided that they determine an intersection of the constructional planes which is not too oblique to afford a sharp result. Nor need the lines of sight be horizontal; we may view an outcrop from above or below, at measured angles of depression or elevation, and hence the method finds application also in the interpretation of oblique aerial photographs (Wallace, 1950), but before proceeding to illustrate this, a small further development of technique is needed.

The method of representation used thus far is *cyclographic*; each structural plane is represented in the stereogram by its great circular trace. The attitude of a plane in space, however, is equally well defined by the orientation of its normal. The two antipodal points in which this normal cuts the sphere are the *poles* of the plane, and if we project stereographically the pole lying on the lower hemisphere we are developing a *polar projection*. In Fig. 25 a structural plane strikes N–S and dips E at 60°; projecting its pole from the zenithal point *P* as usual we obtain a point (Fig. 26) lying on the WE principal diameter of the stereogram 30° inwards from the primitive, and therefore 90° from all points in the cyclographic trace of the plane. Fig. 26 shows both the cyclographic and polar projections of the plane of Fig. 25. For the present we shall continue to use this double representation but as we become accustomed to polar projection, it will often be advantageous to represent all structural planes by the projection of their poles alone. (Strictly only such polar projections are stereograms in the formal sense, since the crystallographer uses polar projection exclusively for the representation of crystal planes; see, however, Fisher, 1952, for a different point of view.)

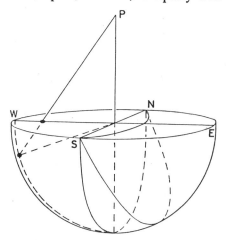

Fig. 25 Cyclographic and polar projection of a plane.

Reverting now to the example of a distant outcrop, Fig. 21, suppose that on viewing *PQ* from the east we occupied an elevated station, so that the line of sight was directed downwards at an angle of 14° depression below the horizontal (Fig. 27). *PQ* is now seen as a projected outcrop on a plane *tvQ* at right angles to the line of

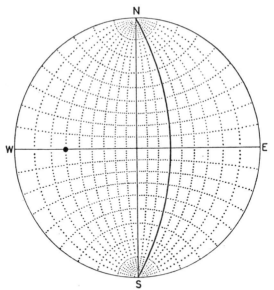

Fig. 26 Stereogram of an east-dipping plane in both cyclographic and polar representation.

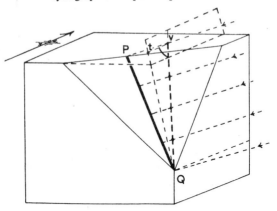

Fig. 27 The outcrop of Figs. 21-3 viewed from an elevated station.

sight (hence dipping east at an angle of 76°). The measured angle tvQ is the angle of pitch of the projected outcrop on this plane. In Fig. 28 the plane tvQ has been plotted stereographically and the angle of pitch (68° southerly) measured along it as the arc ST by means of the graduating small circles. To project the plane PvQ, containing the line of sight and passing through the outcrop PQ, we first insert on the projection the point X, the pole of the plane tvQ (compare Fig. 26) 14° inwards from the primitive. The direction

INTERSECTING INCLINED PLANES 19

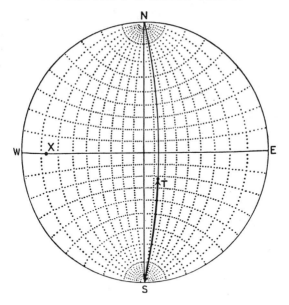

FIG. 28 Stereogram of the observation made in Fig. 27.

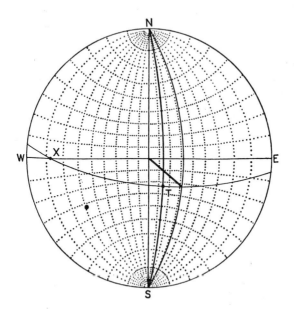

FIG. 29 Determination of the outcrop using the observation in Fig. 27.

of the line of sight is therefore from the centre of the sphere downwards through point X on the lower hemisphere. By revolving the tracing until X and T lie above a common great circle we can trace the required projection of the plane PvQ. Somewhere on this trace lies the pole of the actual outcrop PQ and by combining a second observation the outcrop is located. In Fig. 29 we have used the original first observation of Fig. 22, giving finally, of course, the same result as we reached in Fig. 24.

Lineation on inclined planes of section

Linear structures in laminated or schistose rocks can generally be examined directly in the field by observation and measurement of lineations exposed on the planes of lamination or schistosity. In a hornblende-schist, for example, lineation is usually due at least in part to the preferred orientation of acicular crystals of hornblende; the greatest length of these lies actually in the planes of splitting so that the *lineation* which we examine in the field is in the direction of the actual *linear structure* of the rock itself. With some rock types, however, the problem cannot be approached so directly. Particularly is this true of massive igneous rocks showing flow-banding or linear structure ('rodding') and of foliated but unfissile gneisses. In such rocks the structures can only be examined as lineations on inclined planes of section provided by natural outcrops, glaciated surfaces, joint-planes, etc., or developed by cutting or grinding in the laboratory. These variously-oriented sections must be fitted together to give a true picture of the structure in three dimensions, and stereographic projection offers the quickest route towards an answer.

As a simple example, the following five observations of lineations were made on associated outcrops:—

Plane	Strike	Dip	Pitch of lineation
1	220°	21° south-easterly	76° southerly
2	295°	32° north-easterly	12° north-westerly
3	80°	32° south-easterly	76° south-easterly
4	325°	40° south-westerly	44° north-westery
5	168°	48° south-westerly	86° south-westerlly

In Fig. 30 the poles of the five lineations have been plotted, and in Fig. 31 the tracing has been revolved above the net to show that the five poles can be brought to lie approximately above a common great circle. The observations could therefore be explained if the rock possesses a planar structure striking 310° and dipping 54° south-westerly. This simple determination does not *prove* the presence of such a planar structure within the rock but it does prompt us at once to grind a surface on an oriented specimen parallel to this plane to verify or disprove the suggestion.

If no such simple result emerges, the lineation may well be due to an inherent linear structure or rodding. To check the feasibility of this suggestion and to determine the orientation of the structure if such is present, we may follow the procedure suggested by Lowe (1946).

If the rodding is idealized as a series of mutually parallel cylindrical structures in the rock, the lineation seen on any inclined plane of section follows the longer axes of the elliptical oblique cross-sections of the cylinders (Fig. 32). The plane of the exposure in this figure strikes 225° and dips 55° south-easterly; the pitch of the lineation $L'L$ on it is 60° southerly. If we erect the normal $P'P$ to the lineated plane of observation and pass a constructional plane through

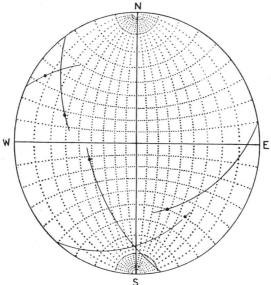

Fig. 30   Stereogram of five observations of lineations.

$L'L$ and $P'P$, then the required axis $X'X$ of the cylinder must be contained in this plane (called by Lowe an N-plane). On the stereogram (Fig. 33) the trace of this N-plane passes through P, the pole of the plane of the observations, and L, the pole of the lineation; hence somewhere on the great circle PL lies the required pole X of the linear structure. In Fig. 34 a second N-plane has been constructed from the observation that on a plane striking 285° and dipping 60° southerly the lineation pitched 58° south-westerly, and the intersection of the two N-planes defines X as representing a linear structure trending W 13° N and plunging 60° westerly.

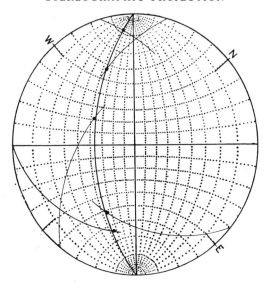

FIG. 31　Stereogram of Fig. 30 revolved until the five poles lie along a common great circle.

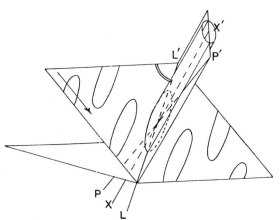

FIG. 32　Idealized lineation as oblique sections of a cylindrical linear structure (after K. Lowe).

In applying this construction in practice, a number of observations on an associated group of outcrops will usually be combined. From n observations the number of solutions is

$$\frac{n}{2}(n-1)$$

In Fig. 35 seven observations are combined. Only the N-planes are drawn, the poles of the associated lineations being marked by crosses. The seven planes yield twenty-one intersections (marked by dots, but some of the intersections coincide). The scattering no doubt arises partly from original difficulty in determining accurately the pitch of an ill-defined lineation, partly from obliquity of intersection of the N-planes and partly from actual changes in the orientation of the structure between one outcrop and another, but Fig. 35 leaves no doubt that all the observations are probably to be explained in terms of a linear structure plunging north-easterly at about 25°.

These two simple examples will serve to illustrate the procedure, which may readily be applied to more complex fabrics such as a

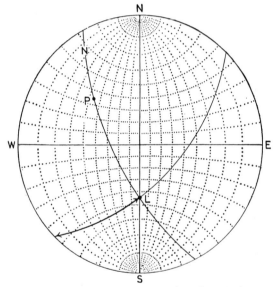

FIG. 33   Stereographic construction of an N-plane.

combination of planar and linear structures; Clark and McIntyre (1951a) work several such examples in detail. It remains, perhaps, to restate that this line of approach is seldom necessary when studying clearly-defined fabrics in well-exposed and 'workable' rocks, but finds its special application to 'streaky granites' and the like.

Joint- and fault-systems; the 'strain ellipsoid'

In a study of the structural relationships of sets of complementary joints or faults one of the immediate aims is to effect a correlation between the orientation of the fracture-systems and the stresses to

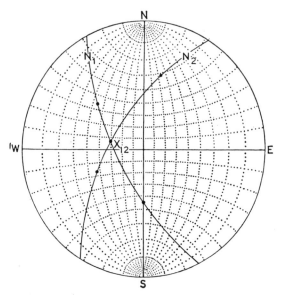

FIG. 34 Stereogram of two intersecting N-planes.

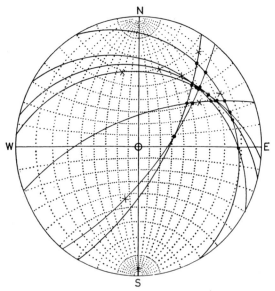

FIG. 35 Seven N-planes intersecting to determine a linear structure.

which the rock has presumably been subjected. This attempted correlation is often referred to as 'orientation of the strain ellipsoid'. We are not concerned here with particular theories of rock failure (see, for example, Nevin, 1949, p. 32) but can accept certain general propositions. Passing through any particular point within a body subjected to stress there are always three mutually perpendicular planes across which there is only normal pressure (or tension). The normals to these three planes are the three *directions of principal stress*; in the general case the values of the three principal stresses will be unequal, giving axes of *greatest*, *intermediate* and *least* principal stresses. In theory, tangential stress is a maximum across two planes which intersect in the intermediate axis and lie at 45° on either side of the other two principal axes. It is found, however (Anderson, 1951, p. 11), that 'the planes of faulting in any rock, instead of bisecting the angles between the directions of greatest and least pressure, will deviate from these positions so as to form smaller angles with the direction of greatest pressure' (Fig. 36). What we actually observe in the rock is the permanent *strain*; it is the strain to which our geometrical analysis relates, and ensuing attempts at correlation between observed strains and inferred stresses lead us into a field of still highly controversial theory.

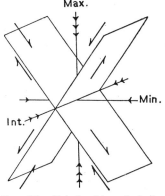

FIG. 36 Relationship of fault-planes to the direction of greatest pressure.

An instructive example of the kind of analysis possible is found in one of our earliest references (Bucher, 1920). In a vertical cliff exposure, trending NNE, of massive sandstone on Mine Fork, Magoffin County, Kentucky, Bucher observed a system of intersecting joints falling into two complementary sets (Fig. 37). The *apparent hades* were measured on the cliff face and the strikes on the level top of the cliff:—

	Set I	Set II
Apparent hade	27° northward	35° southward
Strike	N 78° W (282°)	N 27° W (333°)

After plotting these observations on a stereogram (Fig. 38) we can draw the great circular traces of the two directions of jointing, intersecting at *B*. (Bucher in his original paper projected the upper hemisphere but we adhere to our accepted convention.) If the joints are accepted as parallel to planes of maximum tangential stress then

OB is the direction of the axis of intermediate principal stress. The other two principal axes, lying in a plane at right angles to OB, will emerge on a great circle of which B is the pole, and its trace can be drawn. The dihedral angle between the shearing planes is measured along this great circle (by revolving the tracing until the circle lies along a meridian) as 72°; if we bisect this arc at the point A we obtain a possible position for the pole of the axis of greatest principal stress, it being determined as the greatest stress because it lies in the acute angle between the shearing planes. 90° away from A along the great circle, at A', emerges the third axis, of least principal stress; in this instance it lies only 3° from the horizontal, and Bucher suggested that the stress in this direction was a tension associated with the crest of a minor anticline developed in the sandstone.

When the possible orientation of the three principal axes has been determined along the lines illustrated in the above example, the orientation of further significant planes can be readily deduced.

Fig. 37 Exposure of joint-systems on a cliff face (after Bucher).

Where, for example, the least principal stress is a tension it is probable that open tension cracks (perhaps later mineralized to form gash veins) may develop along planes normal to this axis. The ac cracks and joints of the structural petrologist are believed to lie at right angles to the intermediate axis B. If only certain data are available we can make reasonable assumptions in order to locate the remaining significant directions; thus if the attitude of one set of shear planes is known and also the nature of the movements associated with them (Fig. 36) a tentative orientation can be achieved by assuming that the angle between the known shear and a comple-

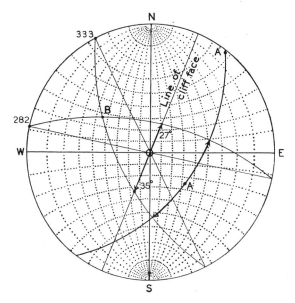

FIG. 38 Stereogram of the joints of Fig. 37.

mentary set is approximately 90° (Cotton and Garretty, 1945, p. 32). See also, Wallace, 1951; Williams, 1958; Badgley, 1959, pp. 206–8; Ramsay, 1962; Price, 1966, Chap. 2, 3.

Distinction of anticlines from synclines

Intersections of different inclinations of the same planar structure in a rock determine fold-structures. It so happens that the only such structures mentioned in this chapter are described as anticlinal, and the beginner may wonder how a syncline is distinguished from an anticline. (This is the geological analogy of the crystallographer's problem of distinguishing *salient* angles, as developed on normal crystals, from *re-entrant* angles such as are frequently found on twinned crystals.) Our projections being concerned only with relative orientation and not with relative position, the answer is that the two kinds of fold yield identical projections, whether cyclographic or polar; any distinction must refer to the actual disposition of the limbs in the field—which is tantamount to saying that we must give the fold its appropriate name.

CHAPTER IV

ROTATION OF THE SPHERE

IN many structural problems it is necessary to rotate the spherical projection, and thus also the stereogram, to some new orientation. In a region of inclined strata, for example, it may be desirable to refer the data to the plane which was horizontal at the time of deposition of the beds. Turning back to Fig. 7 it is clear that, if the sphere is rotated about the line NS, all points on its surface will move along small circles centred at N and S. In projection, for rotation

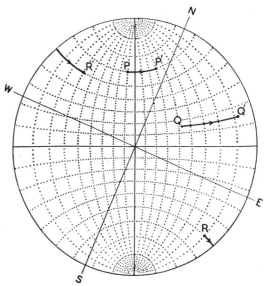

FIG. 39 Rotation of a stereogram through 40° about an axis in the plane of projection.

about the N–S principal diameter, the paths of all points in the projection will lie along the small circular curves of the Wulff net. Hence, to rotate a stereogram about any *horizontal* axis we only need to revolve the tracing above the net until the axis in question lies along the N–S principal diameter and then move all significant poles, etc., through the required angle along the appropriate small circles.

In Fig. 39, for a rotation through an angle of 40°, about an axis trending N 25° W, in a sense which raises the right-hand margin of the projection, the pole P will move to P' and the pole Q to Q'.

The pole R lies only a small angular distance below the primitive and thus the first 10° of rotation bring it on to the primitive itself. Further rotation in the same sense causes the radius through R to plunge downwards towards the north-west and thus the path of movement of the pole continues a further 30°, to R', from the antipodal point on the primitive.

The 'problem of secondary tilt'

REFERENCES:—Earle, 1934, p. 22; Fisher, 1938; Johnson, 1939; Hobson, 1943; Bucher, 1944, p. 203; McLaughlin, 1948; Terpstra, 1948.

One of the most frequent applications of this construction arises in the problem of secondary tilt (sometimes called 'the problem of two tilts'), involving the analysis of structures below an angular unconformity. A lower series of beds, I, of determined strike and dip, are overlain by an upper series, II, with angular discordance;

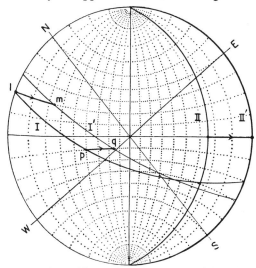

FIG. 40   Cyclographic solution of a problem of secondary tilt.

on the assumption that the bedding of series II was horizontal at the time at which they were laid down it is required to determine the attitude of beds I at that time, before the imposition of secondary tilt.

In Fig. 40 the dip of beds I is 63° on a bearing 240°, below beds II dipping 25° on a bearing 130°. To restore beds II to horizontal, the projection must be rotated through 25°, in the sense shown by the arrow, about the strike of the upper beds—trending N 40° E. To construct the original position of beds I, any two convenient points on its trace are moved 25° in the same sense; in Fig. 40 the point 1 moves 25° from the primitive to m and the point p moves 25° to q, whence the great circle through m and q gives the required original attitude of beds I. After revolving the tracing paper the original dip is read off as 74° on a bearing 248°.

The cyclographic representation of Fig. 40 is probably the easiest for a beginner to visualize. With experience, however, the polar projection is used instead to achieve even greater simplicity. In Fig. 41, as the pole of the beds II moves to the centre of the projection at II′, so the pole of I moves to I′, whence the original attitude of beds I is read off as before.

It will be appreciated that such a solution involves only a few seconds' work and entails merely a common-sense application of elementary understanding of stereographic projection, in contrast with the necessity of manipulating an elaborate trigonometrical formula or of following through a detailed geometrical construction

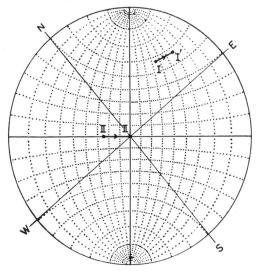

FIG. 41   Polar solution of the problem of Fig. 40.

(Fig. 42, a geometrical solution of the same problem, stands in remarkable contrast with Fig. 41). The gain in simplicity and speed of a single solution by stereographic methods is multiplied effectively in problems involving the conversion of dozens or even hundreds of such measurements. A typical example of its application is afforded by Wager and Deer, 1939, p. 51, where the authors required to eliminate from the structure of an intrusion, as revealed by fluxional banding, the effect of a post-consolidation flexure. It is equally valuable in such a problem as a regional study of false-bedding in inclined strata (e.g., Brett, 1955; Pettijohn, 1957). See, however, p. 35.

A related problem of secondary tilt is encountered in the interpretation of structures revealed in the drill core from a hole which has deviated from the vertical (Johnson, 1939). The apparent dip of a bed as measured in the core has been affected by a secondary tilt expressed by the dip of the cross-section of the core. To interpret a single core in this manner, however, it is of course necessary that the drawn core should be correctly oriented. Various methods of achieving this have been investigated, such as marking the core mechanically before withdrawal, restraining the drill-rods from rotating during withdrawal, or orientating by reference to residual magnetism (McClellan, 1948; Zimmer, 1963). Reorientation of a magnetic core from an inclined borehole involves a twofold application of the solution of secondary tilt described above (Terpstra, 1948). Such procedures, however, have not proved entirely reliable, and most core interpretations involve the study either of unoriented cores from two or more non-parallel boreholes of known attitude (p. 42) or of a dipmeter log (p. 51).

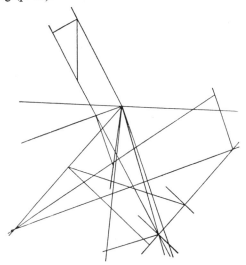

FIG. 42.—Geometrical solution of the problem of Fig. 41.

Application of tilt to a fold

REFERENCES:—Challinor, 1944; Hobson, 1945.

Beds involved in problems of secondary tilt may well be folded. In particular, the series below the unconformity may be folded and it may be required to determine the attitude of the folds at the time when the upper series were laid down horizontally above them. As Challinor (1944, p. 97) comments 'It seems to be very generally assumed that the tilt causing the regional pitch in a series of pitching [plunging] folds was given strictly in the direction of that pitch.

However . . . a fold tilted obliquely will acquire a pitch (or an added pitch) by no means in the direction of the tilt.'

As an example we may consider a fold developed below a plane of unconformity dipping 22° due east. The fold proved to be symmetrical, with a dip on either limb of 35°, but the strikes on the two limbs of 116° and 160° respectively corresponded to an axial plunge of 15° south-easterly. In Fig. 43 the cyclographic traces of the bedding-planes are drawn as a help towards visualizing the fold. To work the problem we insert the poles at *a* and *b*. The pole *U* of the plane of unconformity must be moved by rotation to the centre of the stereogram, so that *a* moves through 22° along a small circle

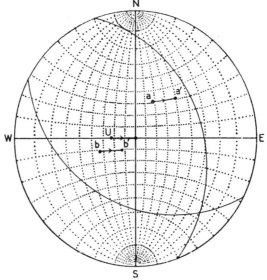

FIG. 43   Stereogram of secondary tilt of a fold.

to *a'* and *b* through the same angle to *b'*. The fact that *a'* and *b'* lie on a diameter of the primitive reveals that before the secondary tilt the axis of the fold was horizontal. The south-easterly trend of the axis has been scarcely affected by the tilt, but the asymmetry of *a'* and *b'* about the centre shows that the fold was originally markedly asymmetric, with dips on either limb of about 16° and 49° respectively; the symmetry of the fold in its tilted attitude is a purely accidental result of the geometry of the movement.

Lineation on folds

Thus far we have been concerned only with projection of the lower hemisphere. In some types of problem it is convenient for easier visualization (though not essential to the working) to project also traces and poles which lie on the upper hemisphere. On p. 2 it was observed that to project the upper hemisphere from the zenithal point *P*, in conformity with the scheme of projection adopted for the lower hemisphere, would extend the projection inconveniently beyond the primitive. We therefore adopt a device long used by crystallographers, and project the upper hemisphere from the point *P'*, the nadir of the sphere, antipodal to *P* (Fig. 44). To indicate that the projection-point has been changed, and that certain of the traces and poles in the completed stereogram refer to the upper hemisphere, the corresponding cyclographic traces are inserted as broken lines and the poles as open rings instead of dots. In Fig. 45 the full trace and the black pole are the cyclographic and polar projections of a plane dipping 24° due east as customarily projected; the broken trace and the open ring are the corresponding projections of the extension of the plane through the upper hemisphere. (Since crystallographers customarily project the upper hemisphere, broken curves and open rings in their stereograms conventionally represent projections of the *lower* hemisphere.)

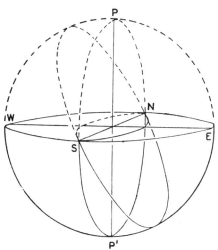

FIG. 44   Projection of a great circle from the upper hemisphere.

The study of lineations on folds affords an example of a problem in which this device is helpful (though unnecessary to the experienced worker). If lineation is visible on both limbs of a fold, the folding may have been impressed upon a previously-lineated plane surface; on the other hand, the lineation may have developed contemporaneously with the folding, or it may possibly have developed subsequently to the folding. A study of the geometry of the situation may suggest which of these possibilities offers the most likely explanation.

On one limb of a fold in pelitic schists, dipping ESE at 25°, a lineation was measured pitching south-easterly at 70°. The axis of folding being horizontal, what would be the trend of lineation on a steep limb dipping 75° in the opposite direction, if the fold has been impressed on a plane surface previously lineated? In Fig. 46 L is the pole of the lineation projected in the usual manner, and L' is the pole antipodal to L. On folding the surface about a horizontal axis, the *angle of pitch* of the lineation *is unchanged*. Hence L' moves over the upper surface of the sphere along a small circle, down to the primitive, and back over the under surface to L''. From the position of the pole L'' on the projection the required trend and plunge of the lineation on the steep limb are determined. If these agree reasonably closely with an observed lineation, the most plausible explanation of the facts lies in the assumption that the development of the lineation preceded the folding movement; since the

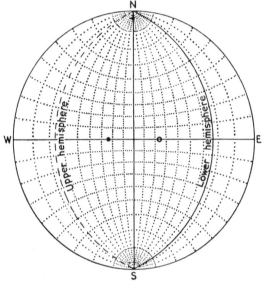

FIG. 45 Stereogram showing the complete projection of a great circle, in both cyclographic and polar representation.

lineation is oblique to the axis of folding it is improbable that the folding movements are closely related genetically to the development of the lineated fabric in the schists.

This illustrative example deals with asymmetric folding but not with overfolding. It is clear, however, that no additional difficulties will arise if the plane passes through the vertical during folding to give an overturned middle limb; L'' continues to travel along its small circle, across the principal diameter of the net (demonstrating the fact of overfolding), to give a final position indicating a lineation plunging north-easterly. If, however, the axis of folding itself plunges below the horizontal it is no longer true that the angle of pitch of the lineation on the surface is preserved unchanged during the development of the fold, and we reserve this most general case for consideration later (p. 40) after we have studied the problem of rotation about inclined axes.

This simplification by restriction to rotation about a horizontal axis holds for the whole of the work in this chapter. The solution of the problems of secondary tilt offered in the earlier paragraphs, for example, assumes that the tilt has taken place about the present strike of the planar structure. The simple solution has been described as geometrically impeccable but geologically unreal. The extent to which this criticism might be valid depends very much upon the scale, and the nature of the structures being studied. To assess the original attitude of a structure in an area of regional tilting our procedure is

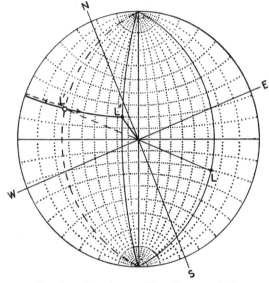

Fig. 46 Stereogram of lineation on a fold.

perfectly valid. Pettijohn, for example, used it in a study of crossbedded quartzites (1957) and wrote of the assumption that the azimuth and inclination of the cross-bedding thus corrected are essentially the same as at the time of deposition 'that this is valid, despite some theoretical objections, has been demonstrated . . . in the Harlech Dome'. If, however, we are concerned with the 'restoration' of a complex region, in which there may be folding about plunging axes, or repeated folding, probably accompanied by shearing and distortion, a different situation clearly arises. The phenomena can still be studied with the help of stereographic projection (see, for example, Ramsay, 1960, 1961), but further techniques are involved which we are about to develop in the next chapter.

CHAPTER V

ROTATION ABOUT INCLINED AXES

IN many structural problems it becomes necessary to rotate the sphere about an axis inclined to the plane of projection. During such movement, any point on the sphere will move along a small circle centred at the pole of the axis of rotation. Since *all* circles drawn on the surface of the sphere project stereographically as circular curves,

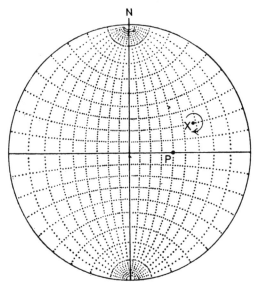

FIG. 47   Stereogram showing pole of an oblique axis of rotation, X.

the path in projection can be fairly readily constructed, but separate constructions would be necessary for each pole to be moved about a given axis and a whole new family of constructions for each new axis of rotation. It is very much simpler to make use of the small circles, centred about the N and S poles of the principal diameter, which are printed on the net. Hence to achieve rotation about an inclined axis we first apply an *auxiliary rotation* to the sphere such as to bring the inclined axis in question into a horizontal position; the rotation involved in the problem is then effected, and the auxiliary rotation finally reversed to bring the sphere back to its original orientation.

This procedure may sound rather involved when thus described in words but it is readily understood in practice. It is, moreover, the foundation for the solution of a number of types of problem, of wide applicability in the fields of structural and mining geology, which lead to the greatest complexity when approached by geometrical or plane trigonometrical methods. It will therefore be worth while to follow through the successive steps in detail. In the stereogram, Fig. 47, P is the pole of a given plane and X that of an axis about which it is required to rotate P through an angle of 60° in the sense shown by the arrow. We proceed in steps:—

1. Mark the north direction on the tracing paper on the margin of the primitive. Revolve the tracing paper until X lies on the W–E principal diameter (Fig. 48). To bring the axis X up to the horizontal the pole X must be rotated through 30° to X^1 and during this rotation P moves 30° along the appropriate small circle to P^1.

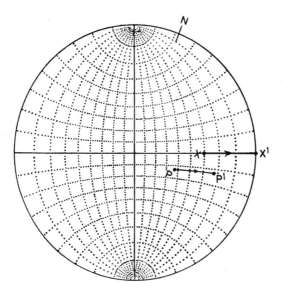

Fig. 48   Stereogram of Fig. 47 revolved to bring X on the W–E principal diameter of the net.

2. Revolve the tracing paper until X^1 lies on the N–S principal diameter of the net (Fig. 49). On rotating the sphere about this diameter through the required angle of 60°, bearing in mind the required sense of rotation and the fact that we are projecting the lower hemisphere, P^1 moves 60° along a small circle to P^2.

3. Revolve the tracing paper through 90° (Fig. 50) to return to the position of Fig. 48. The axis X^1 is now restored to its original position X by reversing the auxiliary rotation of 30°, under which rotation P^2 moves to P^3.

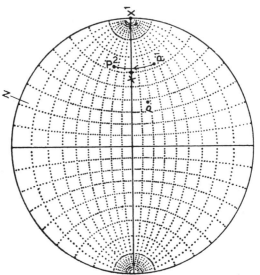

FIG. 49 The same projection revolved to bring X^1 to the centre of the small circles on the net.

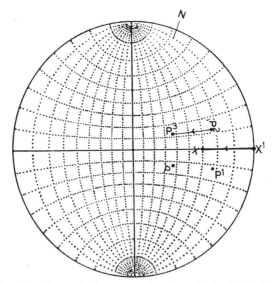

FIG. 50 The projection returned to the attitude of Fig. 48.

4. Revolve the tracing paper to restore the N mark to its correct position (Fig. 51). The final position of P^3 can now be read from the net.

In scissor-movement during faulting the tilting movement of one block relative to the other (as distinct from any displacement due to the faulting) is equivalent to rotation about the normal to the fault-plane; if the fault hades from the vertical the axis of rotation is inclined to the horizontal at an angle equal to the hade of the fault.

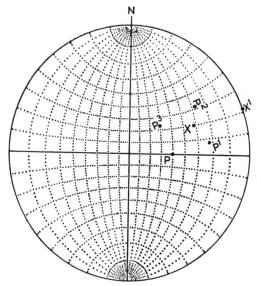

Fig. 51   The N-index finally restored to its original orientation.

In Fig. 52 a fault hades 20° due east, and the scissor-movement is equivalent to a tilt downwards towards the south, through 30°, of the eastern block. In Fig. 53 the poles A and B represent the limbs of a symmetrical horizontal fold in the undisplaced western block; what is the attitude of the fold in the eastern block after faulting? Since the fault hades east at 20°, we rotate its plane into a vertical attitude by an appropriate auxiliary rotation of this amount, thus bringing the axis of rotation into the plane of projection. By this auxiliary rotation A is moved to A^1 and B to B^1. The rotational scissor-movement through 30° moves A^1 to A^2 and B^1 to B^2 (to utilize the printed small circles of the net the tracing paper is, of course, temporarily revolved through 90°), and finally the removal of the auxiliary rotation moves A^2 to A^3 and B^2 to B^3. From the position of these poles, by inserting the corresponding cyclographic traces, we draw a complete description of the attitude of the fold in the tilted block—the fold is now markedly asymmetric about a vertical plane, the trend of the axis has been changed only a few degrees, but there is a steep south-easterly plunge. See also Ho, 1957.

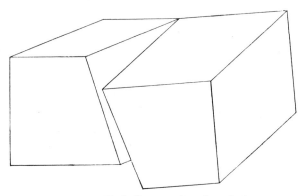

FIG. 52   Block-diagram of a scissor-fault.

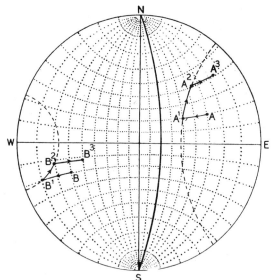

FIG. 53   Stereogram of the effect of scissor-faulting on a fold.

We can next complete our study of lineated folds by considering the general case of the folding of a lineated plane about a plunging axis. On the upper limb of a fold, dipping 20° on a bearing 112°, a lineation pitches south-easterly at an angle of 35° within the plane

(Fig. 54); it is required to determine, for comparison with a field-observation, the attitude of the same lineation when turned over on the overfolded middle limb dipping 60° due south. The cyclographic traces of the limbs intersect in the pole X of the axis of the fold. This reveals a plunge of 20° below the horizontal (Fig. 55) so that

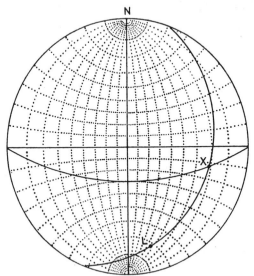

FIG. 54   Stereogram of lineation on a fold.

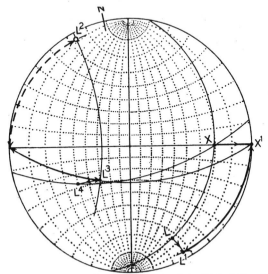

FIG. 55   Rotation of a lineated limb of the fold of Fig. 54.

we must apply an initial auxiliary rotation of this amount to move X to X^1, thus moving L to L^1 (incidentally bringing the upper limb of the fold almost horizontal). The new attitude of the middle limb is also plotted by moving any convenient point on its trace through 20° and drawing the great circle from X^1. The antipodal point to L^1, on the upper hemisphere, is inserted at L^2. On folding the plane over about the *now horizontal axis* the pitch of the lineation as measured from that axis remains unchanged, so that L^2 moves down to the primitive, and back over the lower hemisphere to L^3, along a small circle of radius equal to the arc X^1L^1. Reversing the auxiliary rotation of 20°, by moving X^1 back to X, moves L^3 through 20° to L^4. From the position of this pole, in relation to the N index on the tracing paper, we can determine the trend and plunge of the lineation on the overfolded middle limb and its angle of pitch within that plane.

Interpretation of cores from non-parallel boreholes

REFERENCES:—Mead, 1921; Wisser, 1932; Fisher, 1941; Stein, 1941; Bucher, 1943; Gilluly, 1944; Addie, 1962; Mills, 1963.

The study of concealed structures by means of diamond drilling plays an important part in many structural studies, particularly within the field of applied geology. The extent of the information which can be directly derived from drilling, however, is closely dependent on the nature of the rock revealed in the drawn cores. If a characteristic marker horizon, a key-bed, can be recognized in the succession penetrated by the bores, the depths at which this bed is encountered in three suitably-spaced holes enable the strike and dip of this bed to be readily determined (p. 8). If, however, there is no key-bed which can be recognized in the various cores, the only information afforded by a single unoriented core is the value of the angle made by the bedding-planes with the axis of the core. Though methods have been evolved for determining the original orientation of the core (p. 31), it has been more usual (until the introduction of dipmeters, p. 51) to proceed in such cases by drilling two (or more) non-parallel bores and combining the information which they yield individually. In some of the literature of this method one bore is specified as vertical and the remaining one or more as inclined (parallel bores in this method of

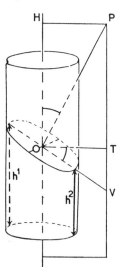

FIG. 56 Block-diagram of a bedding-plane intersecting a cylindrical core.

working of course yield no new information). An initial vertical bore does reveal at once whether or not the beds are dipping steeply; as, however, in practice an initially vertical bore may well deviate as sinking progresses (especially in deep drilling) and as the stereographic method can deal easily with a problem in which all the boreholes are inclined, we shall proceed at once to the general case.

An oblique plane will intersect the cylindrical surface of the core in an ellipse (Fig. 56). We require to measure the angle HOP between the axis OH of the core and the normal OP to the bedding-plane. Passing a plane through OH, OP and the major axis of the elliptical section, it can be seen that $HOP = TOV$ (the complement of the 'core-bedding angle' of some authors). Hence, if we measure h^1 and h^2, the greatest and least distances of the elliptical trace from the terminal cross-fracture of the piece of core,

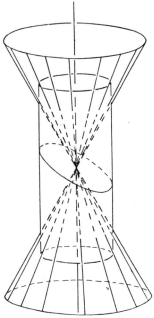

FIG. 57 The double cone of bedding-plane normals corresponding to loss of orientation of the core.

$$\tan HOP = \frac{h^1 - h^2}{\text{core-diameter}}$$

Knowing the angle HOP, but not the orientation of the core, it is clear that the normal to the bedding-plane must lie somewhere on the surface of a double cone with axis OH and semi-vertical angle HOP (Fig. 57). If a similar determination is made for a second hole, with axis OJ oblique to OH, the problem is to determine the orientation of the lines of intersection of the two double cones, giving possible orientations for the normal OP.

In Fig. 58 H is the pole of a borehole plunging 45° on a bearing 220°, and the angle HOP measured 30°; J is that of a hole plunging 60° on a bearing 110°, and JOP was determined as 40°. Though, as we have observed already, it is possible to construct the small circles, in which the cones intersect the sphere of projection, in this orientation (and some authors use this procedure) we shall again find it advantageous to rotate the axes of the cones into the plane of projection and make use of the printed small circles of the net (Fig. 59). Hence, after marking the index N on the tracing paper, we revolve it until H and J lie on a common great circle (Fig. 60), and then rotate the projection through an auxiliary angle to bring this circle into the plane of projection, H thus moving to H^1 and J to J^1. It is then but a moment's work to trace the small circles of radius 30° about H^1 and its antipodal point (Fig. 61) and those of 40° about J^1 and its antipode (Fig. 62). These small circles intersect in the points P and Q, both possible solutions to the problem taken thus far. To

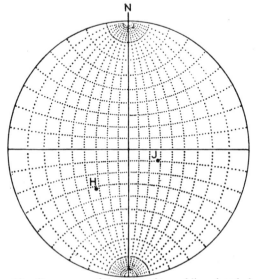

FIG. 58   Stereogram of the poles of two oblique boreholes.

determine the actual orientations of P and Q the initial auxiliary rotation is applied in reverse sense (Fig. 63), bringing P to P^1 and Q to Q^1, and the tracing paper finally revolved to restore the N index to its original position.

There can be little doubt that this procedure affords the simplest possible solution of our problem.

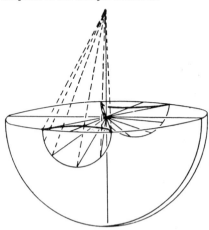

FIG. 59   Stereographic projection of the double cone after the axis of the core has been rotated into the plane of projection.

Once the steps illustrated by the successive figures are understood, very little actual tracing even from the net is really necessary, and Fig. 64 shows the same solution with graphical work reduced

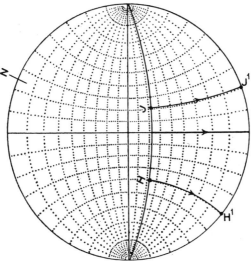

FIG. 60   Stereogram of Fig. 58 revolved until H and J lie above a common meridional great circle of the net

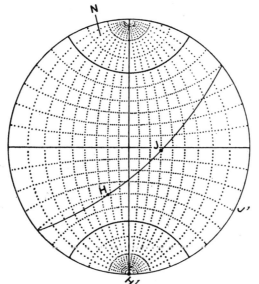

FIG. 61   The stereogram revolved so that the small circles about H^1 can be drawn.

to a minimum. Of the great circle *HJ* only a short arc need be drawn, sufficient to determine the angle of the auxiliary rotation, together with a mark to show the position of the tracing during this operation; the poles *H* and *J* are moved to H^1 and J^1 by counting

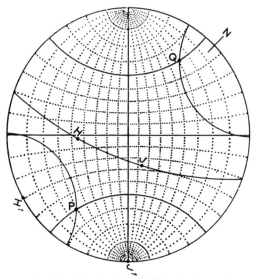

FIG. 62   Tracing the small circles about J^1.

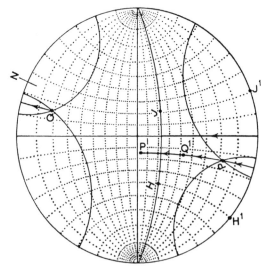

FIG. 63   Movement of the poles *P* and *Q* when the auxiliary rotation is finally applied in reverse sense.

along the graduated small circles without tracing them; short arcs of the small circles about H^1 and J^1 suffice to determine the intersection P, whilst Q is necessarily symmetrically opposite P on the net; finally, P and Q are moved to P^1 and Q^1 by again counting along the appropriate small circles.

The discovery of two solutions which fit the data derived from two non-parallel bores illustrates only one of several possible situations which may arise. The number of intersections of the two pairs of small circles is interdependent on the angles of the cones and the

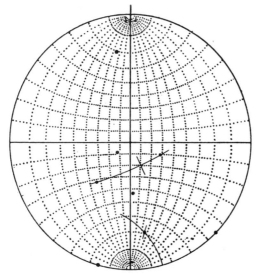

FIG. 64 Stereogram of the construction of Figs. 58–63 as effected in practice.

mutual inclinations of their axes. Fig. 65a illustrates two intersections, which we have encountered above, but in Fig. 65b the small circles intersect at four points. In Figs. 65c and d, owing to the small circles just touching, the number of possible solutions is three and one respectively. (One writer on this subject speaks of two circles 'really intersecting at three points', scarcely a happy reference to the mutual intersections of *two pairs* of small circles!) If the small circles fail to intersect, the original data are clearly inconsistent; either there is an error in determining the plunge of the bore at the point from which the core was taken or else the attitude of the bedding has changed between the two cores derived from points too widely spaced.

The selection, from amongst two, three or four possible solutions. of the single correct answer may sometimes be made from general considerations such as a knowledge of the approximate direction of regional strike. It is usually readily made if information from a third borehole, L, is available. The pole L is inserted on the original projection with H and J; when the final positions of the intersections, P^1, Q^1, etc., have been determined, the angular distances LP^1, etc., are read from the net by revolving the tracing until the pairs of poles lie in turn above common great circles. By comparing the two, three or four values with that of the angle LOP measured on the third core a single selection can usually be made, though in unfavourable cases it may be necessary to resort to a fourth hole. Closeness of agreement between the data from multiple bores is affected by the possibilities of error or divergence noted above, but examination of the data from the first two bores may suggest a critical attitude for the third hole before drilling is undertaken.

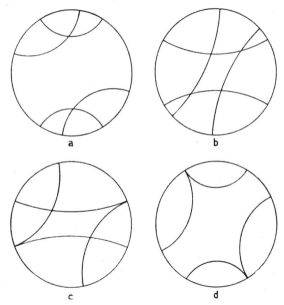

FIG. 65   Intersections of small circles (representing double cones of bedding-plane normals).

Cyclographic projection

Despite the simplicity of polar stereographic projections, it has been argued from time to time (e.g., Mills, 1963) that it is easier for most geology students to visualize the attitude of a plane when it is represented cyclographically. We therefore proceed to examine briefly here the application of this mode of projection to borehole problems.

ROTATION ABOUT INCLINED AXES

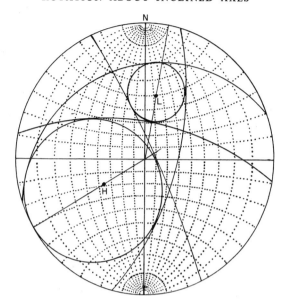

FIG. 66 Cyclographic solution for non-parallel boreholes. Four solutions from two boreholes.

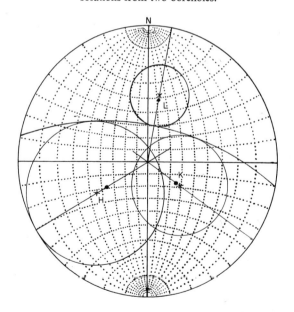

FIG. 67 Unique solution by adding information from a third borehole to Fig. 66.

Just as the normal to a bedding plane, in the polar solution, must lie on the surface of a double cone with semi-vertical angle HOP (Figs. 56, 57), so also the bedding plane itself must be tangent to a cone the semi-vertical angle of which is 90°—HOP. In Fig. 66 two inclined boreholes are projected at H and L, and about these poles are drawn small circles of radii corresponding to the complements of the measured angles between the core-axis and bedding-plane normals. The required cyclographic trace of the bedding plane must touch both circles, and four possible solutions have been constructed graphically by revolving the tracing above the net. In Fig. 67 a further small circle has been drawn, in accordance with data derived from a third borehole, K, and this determines a unique solution.

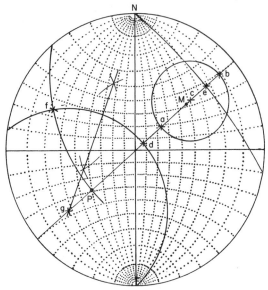

FIG. 68 Small circle constructions for cyclographic solutions.

Small circles of this kind, the centres of which lie within the primitive, cannot, of course, be traced from the net, but must be constructed individually. Fig. 68 illustrates the construction of a small circle of radius 25° about a centre M, the projection of a borehole plunging 35°. The tracing is revolved until M lies on the NS (or WE) diameter of the net, and angular distances of 25° are counted off from M, towards the centre to give the point a and towards the primitive to give the point b. The centre of the required circle in projection must then lie at the point c, half-way between a and b (it does not, of course, lie at the point M which is the projection of the centre on the sphere).

For a small circle of radius 60° about a point P, marking a plunge of 40°, one extremity of the diameter is readily located at *d*, as before. The other extremity, however, lies beyond the primitive, and projects at *e*, on the opposite side of the projection, if we follow our customary procedure. The projection could be extended beyond the primitive, by a device familiar to crystallographers, but a simpler construction can be used to locate the required centre *g*. The tracing is revolved to bring P on the WE diameter, and the meridian through P enables the angular distance P*f* of 60° to be counted off to locate *f*, a point on the required small circle. The chord *df* is then bisected to give the required centre at *g*. The remaining part of the circumference of the small circle, on the opposite side of the projection, of course passes through *e*. In the present instance the radius of this in projection is quite large, and the arc can only be drawn conveniently with the help of a flexible ruler or a draughtsman's curve.

The preceding paragraphs bring out the tiresomeness of the necessity to construct small circles centred within the primitive. Still more troublesome is the position which arises if we use an equal-area net (p. 60). Since the trace in projection is no longer circular, several points on the circumference must be determined before it can be traced (Turner and Weiss, 1963, pp. 57-8). These considerations seem to the author to add up to a very good reason why the student should learn to appreciate the significance of polar representation as early as possible. Even if the final determination needs to be presented in relation to other data in an equal-area projection, the intermediate constructions are most easily carried out on a Wulff net.

Dipmeters

In view both of reduced expense and of increased speed of production of information it is highly advantageous if reliable information about subsurface dips can be extracted from a single borehole. Attempts to design a 'dipmeter' by means of which a hole could be surveyed for this purpose began in the nineteen thirties, but some twenty years elapsed before complete success was achieved. An early type logged three simultaneous profiles of the borehole wall, from which measurements could be made of the vertical displacements of the protruding edges of hard beds (Boucher, Hildebrandt and Hagan, 1950). Other physical properties of the beds were later invoked, such as spontaneous potential and resistivity. A modern instrument carries three spring arms which press three groups of electrodes, 120° apart, against the wall of the hole (Fig. 69). The centres of the electrode groups lie in one plane, perpendicular to the axis of the instrument and hence to the axis of the hole. As the dipmeter is drawn upwards, seven parameters are continuously recorded on a chart—the three correlation curves from the three electrode-groups, the azimuth of no. 1 group, the diameter of the hole, and the inclination of the hole either as the components of inclination in the NS and WE directions or directly as the azimuth and amount of true inclination.

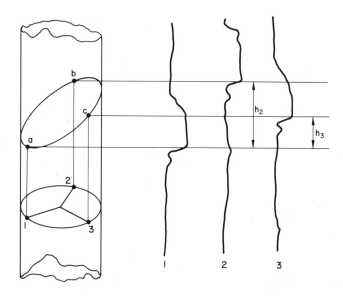

Fig. 69 Diagram illustrating the principle of the continuous dipmeter.

Fig. 69 shows diagrammatically how the differences of height, h_2 and h_3, between the three points a, b, c at which a particular bedding plane is encountered by each of the three electrode-groups 1, 2, 3 are determined from the three correlation curves. From Fig. 70 we can then determine the angles of apparent dip bab' and cac':—

$$\tan bab' = \frac{h_2}{1 \cdot 732\, r} \qquad \tan cac' = \frac{h_3}{1 \cdot 732\, r}$$

where r is half the recorded diameter of the hole at this position. From these two apparent dips, in vertical planes at angles of 30° on either side of an electrode the azimuth of which is known, the azimuth and amount of 'true' dip are determined in the usual manner (p. 7).

This dip, however, is relative to the axis of the borehole, and must be finally corrected for inclination of the hole. If the azimuth and amount of this inclination are recorded directly, as in some types of dipmeter, the appropriate displacement can be applied directly to the pole of the uncorrected attitude of the bedding plane. If the dipmeter records the inclination in terms of the NS and WE components, these must first be projected, in an analogous manner to apparent dips, in order to determine the amount and azimuth of true inclination. For small deviations from the vertical (a modern instrument can in fact be used in holes plunging at low angles) it may be advisable to use the complementary values of the components

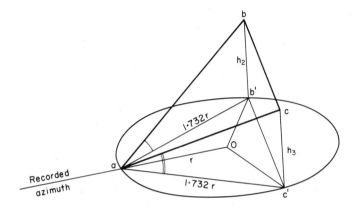

FIG. 70 Interpretation of a dipmeter record.

when plotting, in order to bring the construction to the more open, marginal, area of the net (Badgley, 1959, p. 214. Note that this author uses an upper hemisphere projection when determining the uncorrected dip of the bedding plane).

A dipmeter records azimuths in terms of magnetic north, and the final dip determination must be corrected for the prevailing declination.

CHAPTER VI

STEREOGRAPHIC PROJECTION AS AN AUXILIARY TO FINAL SOLUTION

IN the problems discussed thus far, graphical work on the stereogram has led directly to the final answer sought. In some other types of problem a final geometrical construction may be necessary, the stereographic method proving a valuable auxiliary in determining quickly the angular values needed for this final construction. Two groups of examples of this kind of application will next be considered.

Calculations in fault-planes

REFERENCES:—Sokol, 1927, pp. 198, 207; Morishita, 1938; Beckwith, 1947; Badgley, 1959, pp. 199–202; Haman, 1961; Ragan, 1968, pp. 96–105.

Problems solved completely within the field of stereographic projection are usually such that the geometry can be referred to a single fixed origin, since we are concerned only with *directions* which relate the orientations of planes and lines. In fault-problems we are concerned with displacement of the origin and with the relative *positions* of the displaced structures; displacement cannot be directly expressed on a stereogram, and we proceed most easily by projecting the structures on to the plane of the fault, when the characteristics of the faulting such as the net slip (or resultant path) will at once be evident. The accurate construction of such a projection involves the determination of a number of angles of apparent dip or pitch (Haarmann angles) and we have learned above how readily such determinations are made from a stereogram.

The following example is taken from Billings (1954, p. 448). The fault FF' (Fig. 71) hades 50° (i.e., dips 40°) due south. On the south side of the fault a vein is exposed at A dipping 35° on a bearing 60°, and reappears on the north side at A' with the same strike and dip; a second vein, exposed at B and B', dips 60° on a bearing 300°; we require to determine the net slip. The cyclographic traces of the fault-plane and of the two veins are projected in Fig. 72, the latter intersecting the fault at P and Q respectively. The arc α_A is the pitch of the trace of the A vein on the fault-plane, and in Fig. 71 lines AS and $A'N$ are drawn making this angle with the strike of the fault FF'. Similarly, the value of α_B on the stereogram determines the direction of the lines $B'N$ and BS. The intersections N and S are *corresponding points*, and NS gives the direction of net slip on the plane of the fault and the amount of slip when referred to the scale of the drawing.

AS AN AUXILIARY TO FINAL SOLUTION

We have already seen (p. 39) how the stereographic projection can be used to evaluate the effect of a rotatory component in the movement of the fault. From a projection on the fault plane of the structures on either side, the direction and amount of a hypothetical rotation to account for the observed relationships can be deduced

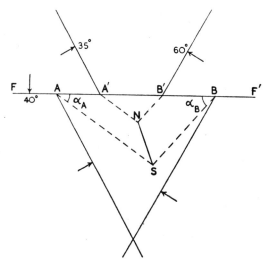

FIG. 71 Plan of faulted veins (after Billings).

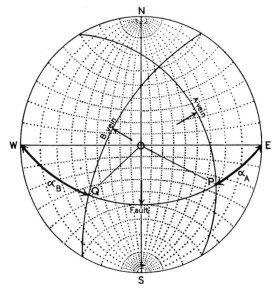

FIG. 72 Stereogram of the fault and veins of Fig. 71.

(though this, without other supporting evidence, may not represent the actual movement, any more than a resultant path need necessarily represent the actual path of movement) and the stereogram offers the quickest route to construction of such a projection. Several examples are discussed in detail by Beckwith (1947) and it is unnecessary here to elaborate the matter further. The reader should consult Wallace (1951) for an interesting application of stereographic projection to the study of the direction of movement on a fault-plane; see also Williams, 1958; Haman, 1961.

Construction of block-diagrams

REFERENCES:—Stach, 1923, 1929; Lobeck, 1924; Johnston and Nolan, 1937; Ives, 1939.

A further field in which stereographic methods are a valuable auxiliary is that of the construction of accurate block-diagrams (or models) of geological structures. A structural plane intersects the faces of the block in lines having an apparent dip (or pitch) within the face which can at once be determined from the stereogram; the procedure for correctly reproducing the determined angle in the diagram depends, of course, upon the method of projection adopted in the drawing. An isometric projection has many advantageous features and has been widely used (Stach, 1922, 1929; Lobeck, 1924, pp. 119–47; Wegmann, 1929a; Johnston and Nolan, 1937). Drawings can be made directly on isometric graph paper familiar to engineering draughtsmen (e.g., the Chartwell Series, obtainable through Messrs. H. K. Lewis & Co. Ltd., 136 Gower Street, London, W.C.1; this paper is scaled in light blue, so that the background will not appear if a diagram is reproduced photographically).

In most diagrams of this kind, the axes are so oriented that the top surface of the block ($ABCD$, Fig. 73) represents a horizontal plane (e.g., ground surface, or a particular level in a mine) and the vertical sides of the block are set NW–SE and NE–SW. Suppose that on such a unit block it is required to construct a plane striking 340° and dipping 30° westerly. On the horizontal top surface of the block Ap represents the true strike of the plane and hence $BAP = 25°$; since, from tables, $\tan 25° = 0.47$ and the side AB has been drawn 2 units in length, we mark off $BP = 2 \times 0.47$, and join PA (cf. Ives, 1939). If tables are not available, or if the diagram is being drawn without the help of isometric graph paper, the direction of the line AP is easily determined graphically as shown in Fig. 74. On the square $AbCd$ the line Ap makes the actual angle of 25° with the side Ab; as the square $AbCd$ is distorted to the rhombus $ABCD$, the point b moving to B, the point p moves to P, giving the required direction AP. For the pitch of the trace PQ on the right-hand face of the block we read from the stereogram (Fig. 75) that the angle of pitch in this plane is 28°; tan 28° = 0.53, so that *CM* is marked off 1.06 along the vertical edge *CF*, and *BM* then represents the slope of the required trace, drawn as a parallel at *PQ*. Similarly, the pitch on the left-hand face is determined from the stereogram as 14°; tan 14° = 0.25, so that *DN* = 0.50 and *QR* is drawn parallel to *CN*. The join *RA*, parallel to *QP*, completes the construction.

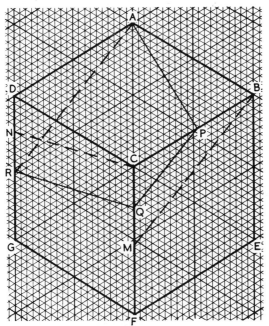

FIG. 73 Use of isometric graph paper in constructing a block-diagram.

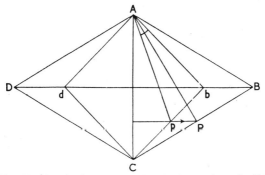

FIG. 74 Geometrical construction for isometric projection.

Another frequently-used projection is based on the conventional clinographic axes of the crystallographer (Fig. 76). The faces here would usually be set W–E and N–S, so that the required angles of pitch are read from the intersections of the cyclographic trace on the principal diameters of the stereogram (in Fig. 75 the values are 28° on the front face and 11° on the side face). The representation of the angles, from the values of their tangents, is accomplished in exactly similar manner to that adopted for the isometric projection and explained in detail above, but in clinographic projection the units of

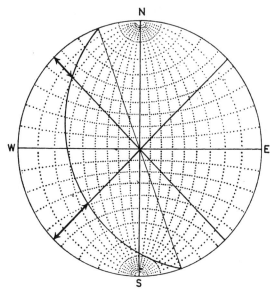

FIG. 75   Stereogram for the construction of Fig. 73.

measurement parallel to the three axial directions are all different and depend upon the direction of view chosen for the construction of the axes (Phillips, 1963, pp. 46–8). To introduce a simplification, various modified projections have been proposed; in Fig. 77 the same plane has been constructed once again, this time on the 'cabinet projection' of Lobeck (1924, p. 190). The vertical and W–E axes are at 90° on the paper and equal units of measurement are used along them, so that the front face of the block is truly square in the drawing and the pitch angle of 28° reproduces as such on the paper. The N–S axis makes an angle of 60° with the vertical axis on the paper, and a unit of measurement along it is one-half of that along the other axes, so that computation from the values of tangents for the construction of lines on the upper and right-hand faces of the block is simple.

AS AN AUXILIARY TO FINAL SOLUTION

Further planes, representing unconformable strata, faults, veins and so forth, are added to the block-diagram after inserting the cyclographic trace on the stereogram, and from their mutual intersections a complete block-diagram even of a complex structure is readily built up in the usual manner.

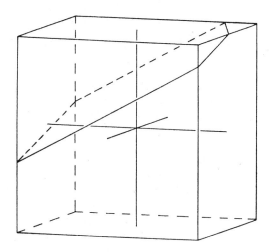

FIG. 76 Block-diagram of the plane of Fig. 73 on clinographic axes.

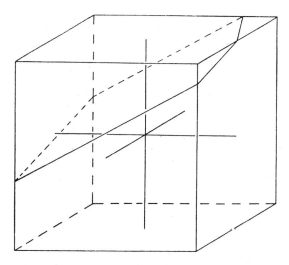

FIG. 77 Block-diagram of the same plane on 'cabinet projection'.

CHAPTER VII

TECTONIC SYNTHESES IN STEREOGRAPHIC (AND RELATED) PROJECTION

REFERENCES:—Seitz, 1924; Wegmann, 1929, 1929 a, 1929 b; Fischer, 1930; Müller, 1933; Sander, 1942, 1948; de Cizancourt, 1947; Rodgers, 1952; Billings, 1954, p. 116; Whitten, 1957; Badgley, 1959; Turner and Weiss, 1963; Ramsay, 1964.

PRECEDING sections have described the application of stereographic projection to the solution of various kinds of specific structural problems. It remains to illustrate the usefulness of the projection in synthetic studies of the tectonics of a region. A geological or structural map, of course, provides such a synthesis but it is not always easy, particularly when a map covers a large area, to view the region as a whole in terms of the structural symbols scattered over the map in their actual geographical locations. A more comprehensive view is obtained if a large number of scattered observations (such as the hades of a series of joints) are brought together about a single origin in a composite diagram. This kind of device has long been familiar, for example, in the 'compass-rose' diagrams of joint- and fault-systems, but a stereographic projection has superior properties—Wegmann wrote in 1929 '*Das Wulff'sche Netz . . . muss die sogennanten Kluftrosen, welche dem Wiegenalter der Tektonik entsprechen, ersetzen.*'

Before entering upon a discussion of some illustrative examples, we must first pay attention to two new considerations which now arise. Firstly, in a synthesis of this kind we are no longer concerned with a few specific directions represented by their individual poles in projection, but with the *grouping* of a large number of measurements —e.g., 200 poles of measured joints—*on the surface of the sphere.* The stereographic method of projecting a sphere does not include, amongst its many advantageous features, that of being area-true; it can readily be seen from the Wulff net that an area bounded by four arcs of 10° projected near the centre of the stereogram is much smaller in projection than the same area of surface of the sphere projected near the primitive. Hence in problems of the kind now under consideration there is considerable advantage in using, in place of the stereographic, a scheme of projection of the sphere which is area-true. The *Lambert equal-area projection* is customarily used, and Fig. 78 illustrates a net of the familiar great and small circles projected on this scheme. Such a net is often called by geologists a *Schmidt net*, after W. Schmidt of Berlin who introduced its use in structural petrology. We need not concern ourselves here with the geometrical basis of this scheme of projection. The net is used for plotting in precisely corresponding fashion to the Wulff stereographic net, and printed versions, either unmounted or mounted, 20 cms. diameter, can be obtained from the sources mentioned on p. 5. (Such a net can equally well be used throughout the work which has been discussed in the earlier sections, but, since the curves which represent in projection the traces of the families of great and small circles are not circular, the equal-area net is not so convenient as the stereographic for graphical work involving cyclographic representation.)

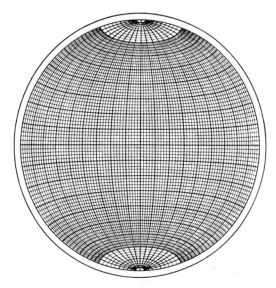

FIG. 78   A Lambert equal-area net ('Schmidt net').

Secondly, in most synthetic diagrams, whether they bring together observations of lineations, fold-axes, etc., or of normals to bedding-planes, joints and so forth, polar representation will usually be employed rather than cyclographic. The fundamental data being the azimuth of each of the linear directions in question and its angle of plunge, little need arises for the traces of meridional great circles of the usual type of net; plotting is easier on a *polar net* (Fig. 79) in which the family of small circles is centred about the extremity of the axis normal to the plane of projection and the great circles pass through this axis. Fig. 79 is a *polar stereographic net*, as used, for example, by de Cizancourt (1947) in his study of joint-systems. (This author extends his projections 40° beyond the primitive, so that more than a hemisphere is represented; there seems to be little advantage in this procedure to anyone thoroughly conversant with stereographic projection.)

By spacing the small circles in accordance with the Lambert scheme of projection the corresponding *polar equal-area net* is constructed (Fig. 80); diagrams figured directly on this net are a prominent feature, for example, in many studies of preferred orientation of pebbles in sedimentary deposits (e.g., Krumbein, 1939). See also Harrison, 1957.

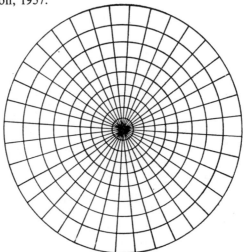

FIG. 79 A polar stereographic net.

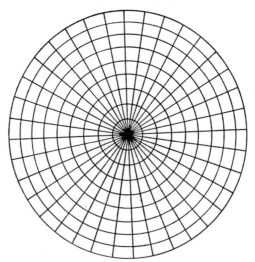

FIG. 80 A polar equal-area net.

TECTONIC SYNTHESES 63

To complete our description of the types of net which the geologist is likely to encounter, it should be mentioned that crystallographers sometimes use the Fedorov stereographic net (Fig. 81) in which all the families of polar and meridional circles are represented, but the greater confusion of lines makes this net less useful in most geological work.

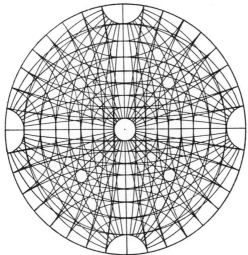

FIG. 81   A Fedorov stereographic net.

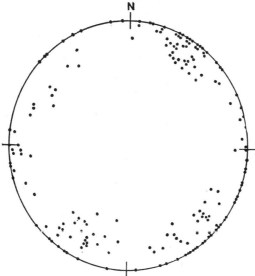

FIG. 82   Equal-area projection of the poles of joints, Start Point.

An equal-area plot of the poles of 185 sea-eroded gullies along joints in mica-schists on the coast west of Start Point, South Devon, is shown in Fig. 82. Qualitatively, the general characteristics of this system of joints are well summarized in this diagram—there is clearly a concentration of poles near the primitive in the north-easterly quadrant with a rather lower one in the south-east, and whilst many of the joints are vertical some depart 30° or more from this attitude. The details of the pattern of joints, however, can be better appreciated if the pole-diagram ('scatter-diagram') is con-

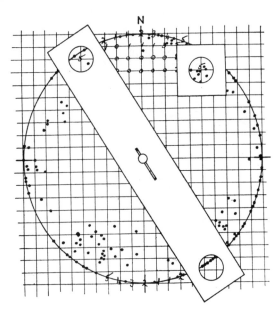

FIG. 83   Counting out the projection of Fig. 82.

toured after the manner customary in studies in structural petrology (Fischer, 1930, p. 20; Müller, 1933; Billings, 1942, p. 116). The percentage concentration of poles over each unit of surface area of the sphere is determined by moving across the projection a circle* of 1 cm. radius cut in a thin sheet of plastic or metal (Fig. 83); in the position shown, nine poles are visible within the area of the counter, and there are 185 measurements plotted on the diagram, so the figure 5 is written on the projection at the centre of the counter. Though the counter may be moved at random, it is more usual to

*The use of a *circular* counter on an equal-area projection is not theoretically justifiable, since small circles project in general as ellipses and not as circles on the Lambert scheme, but the resultant inaccuracy is small and a circular counter is universally used for simplicity.

centre it successively on the intersections of a square grid placed beneath the tracing paper as shown in the figure; Fischer (1930), having plotted his diagram on a polar net, suggests centring on the intersections of the great and small circles of the net. For points close to the primitive, a double-ended counter devised by Schmidegg is used; in Fig. 83 three poles lie within the upper circle, but the 1 per cent area of the projection is completed by the portion of the lower circle enclosing six further poles, so that the figure 5 is again written at the centre of the upper circle. For centres lying on the

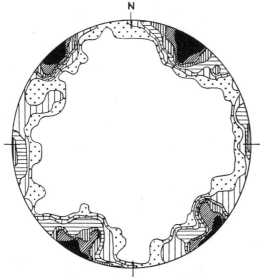

FIG. 84 The projection of Fig. 82 contoured and shaded.

primitive, the same figure is written at both centres (e.g., the figure 2 at N and its antipode). When the scatter-diagram has been thus counted out, 'contour-lines' are drawn around areas of like density of occupation, and the details may be brought out by some form of conventional shading (Fig. 84). (Fischer appears not to have been conversant with the Schmidegg perimeter-counter; the contouring of his figures 15 and 17 (1930) is geometrically incorrect, diametrically opposite points on the primitive not corresponding; it should be clear from an understanding of what such projections represent that any contour reaching the primitive must necessarily re-enter the projection at the antipodal point.)

From Fig. 84 we can read more detail about the joint-system at Start Point; the black maxima indicate two complementary sets of diagonal joints making an angle of about 110° almost symmetrically about the N–S direction, and the minor area of 4 per cent marks a subsidiary set of joints bisecting this angle; the symmetrical placing of the maxima on the primitive shows that the joints, though sometimes hading individually, are statistically vertical.

In studying a fold-system we can most conveniently plot the poles of all the planes of bedding or schistosity of which the dips have been measured (Wegmann, 1929b). If there is regular folding about parallel axes these poles will lie along a great circular arc, which is readily traced on the stereogram. From the position of the pole of

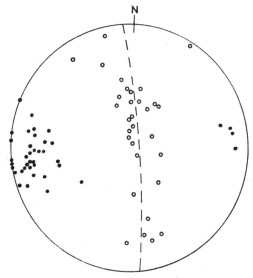

FIG. 85 Stereogram of normals to schistosity and poles of lineations, Start area, S. Devon.

this great circle we can read the azimuth and amount of plunge ('pitch') of the axis; the degree of scatter about the great circle reveals the divergence amongst individual folds from an ideal attitude. The open circles of Fig. 85 record in this manner a number of observations on the schists of the Start area, South Devon, and their disposition is clearly in accordance with folding about nearly W–E axes with a gentle plunge to the west, which is the generally accepted structure of the area deduced from a study of the outcrop of a marker horizon of green schists. To the same diagram we can add further information. The full black circles are the poles of observed lineations on the schists, which are proved by microscopic work to be B-axes of the structural petrologist—they are normal to a monoclinic symmetry plane of the fabric. It is clear from the projection that in the Start area we have the simple connexion between folding and fabric that the *B*-axis of the fabric coincides with the axis of folding.

Turning next to Fig. 86, a similar diagram is shown for a number of observations on Moine schists from the North-West Highlands of Scotland. Here there is no such simple relation; whilst the lineation (which again represents a *B*-axis of the fabric) plunges south-easterly with considerable scatter, the poles of the planes of schistosity show little tendency to group about a great circle of which the lineation is the normal. The relationship of macro-folding to micro-fabric in the

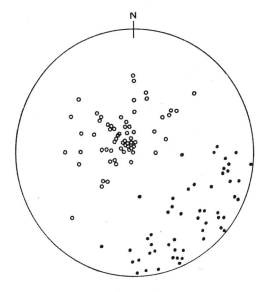

FIG. 86 Stereogram, in the manner of Fig. 85, of observations on Moine schists.

Moine schists is clearly more complex than that of the Start area, and the contrast between Figs. 85 and 86 must have an important bearing on discussions of the metamorphic history and age-relationships of the Moine schists.

Further data can be added to stereographic plots of this kind—poles of joint-planes, poles of elongated fabric elements such as 'stretched pebbles' and so forth. When synthesizing data from a large map-area it is advisable, initially at least, to make separate projections of the data from adjacent smaller areas (Wegmann, 1929b). By superposing these tracings one can detect any systematic change of orientation over the larger area which might well be concealed in the general scatter of a single composite diagram.

These methods have been widely used by structural petrologists and Sander (1942; 1948; in English translation 1970, pp. 133 ff.) developed a systematic notation and symbolic representation. Fabric planes such as those of bedding and schistosity are s-planes, and the normals to s-planes are projected as open circles as we have done in Figs. 85 and 86. The great circle on which these poles tend to lie Sander termed a π-circle, and he represented it in cyclographic projection by a broken curve; the pole of the circle he termed the π-pole. B-axes are projected as full black circles. Instead of plotting the poles of normals to s-planes, these planes may be drawn in

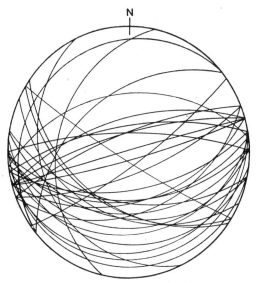

FIG. 87 Cyclographic stereographic projection of the planes of Fig. 85 (β-diagram of Sander).

cyclographic representation as full black curves (Fig. 87). Their intersections determine a number of β-poles, which may be plotted independently as small crosses. Geometrically, the β-maximum corresponds to the π-pole, but as we have seen above β and B may not coincide.

Some confusion has arisen through misuse of these symbols by later writers, and as π and β are 'different symbols for the same element of the fabric' (Turner and Weiss, 1963, p. 155) it has been suggested that use of the symbol π be discontinued. It remains true, however, that β axes are most satisfactorily located graphically from the π-circle (we can call the plot an s-pole diagram) rather than directly from great-circle intersections. The cyclographic representation, as in Fig. 87, is clearly unsuited to more than about thirty planes (Ramsay, 1964); the mass of data now often available can best be handled by a computer (Robinson *et al.*, 1963; Nobel and Eberly, 1964).

Conclusion: speed and accuracy of the method

Emphasis has been placed throughout on the speed with which various kinds of structural problems can be solved by simple graphical work with a stereographic net. From time to time devices have been suggested to augment this further, and they may be worth consideration if much work of the kind is to be undertaken. In the *Kohlschütter* apparatus a net printed in red on a translucent background revolves above a similar fixed net printed in black on an opaque white ground. Wallace (1948) described a similar stereographic calculator from which the answer to simpler problems, such as those of apparent dip, can be read directly without carrying out any graphical construction. Stockwell (1950), when describing the graphical construction of cross-sections of folds by projection down plunge, uses a 'duo-stereogram' from which the bearing and plunge of lines of intersection of planes, taken two at a time, can be read directly from their strikes and dips.

The accuracy obtainable is, of course, related to the size of net used. On the 20 cm. net graduated every two degrees, it is usually easy to work in the laboratory with an accuracy well within the limits which most natural exposures permit in the original data. On a Wulff net, the greatest accuracy can be maintained by working near the edge of the net, where the graduations are more open, so that problems dealing exclusively with low dips may be solved in cyclographic representation and the polar method be used when the dips are predominantly high, or recourse may then be had to a Schmidt net. An additional graphical error is liable to be introduced if a construction should come to depend upon arcs intersecting at a low angle (cf. Wallace, 1950, p. 279) and if possible the original observations should be so made as to avoid this situation arising. If for some particular purpose a specially high degree of accuracy is desirable (an occasion unlikely to arise in routine geological work) very large reproductions of a Wulff net, specially prepared for use in astronomical and navigational plotting, can be obtained. Alternatively, at the cost of an increased consumption of time, the required angular values may be calculated from the stereogram (see Appendix, p. 70).

APPENDIX

CALCULATIONS BY SPHERICAL TRIGONOMETRY

REFERENCE:—Higgs and Tunell, 1959.

FOR the sake of completeness, we add here a brief account of the use of the formulae of spherical trigonometry in the calculation of values on a stereogram. Though, as we have suggested, the degree of accuracy obtainable in the original data of a geological observation will rarely justify the accuracy of such calculation a geologist may occasionally wish to check his graphical solutions. Moreover, the methods which we have discussed have applications far beyond the purely geological field; a stereogram affords the neatest presentation of orientational data in many physical, structural and engineering problems, and calculation from the stereogram avoids the clumsy manipulation of plane trigonometry in three dimensions (see, for example, Sander, 1970, pp. 131–3).

The non-mathematical geologist need not be repelled by the words 'spherical trigonometry' in the title of this appendix (though one writer reaches the conclusion that a particular solution 'probably involves spherical triangles and consequently would be of use to few geologists'!). Provided that one understands the meaning of the sine, cosine and tangent functions of an angle, and can use logarithmic tables of these functions, those formulae required can be manipulated as easily as those appropriate to plane trigonometry.

A *spherical triangle* is delineated on the surface of a sphere by the intersection of three great circular arcs; it has six *parts*, the three *sides* and the three *angles*. Unlike plane trigonometry, however, we are not concerned with the lengths of the sides—the size of the triangle (or the radius of the sphere) is immaterial, and the sides are expressed in circular measurement as the angle which they subtend

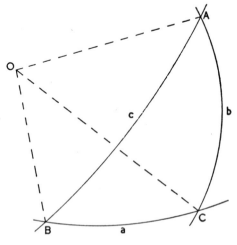

FIG. 88   The parts of a spherical triangle.

CALCULATIONS BY SPHERICAL TRIGONOMETRY

at the centre of the sphere. In Fig. 88 the side a is measured by $\angle BOC$, the side b by $\angle COA$ and the side c by $\angle AOB$, where O is the centre of the sphere. A spherical triangle is determined if any three of its parts are of known magnitude, and formulae are available for calculating the three unknown parts in any possible case. Two important differences from plane trigonometry must, however, be noted; since the size of the triangle is immaterial, a spherical triangle is determined if all three angles A, B and C are of known magnitude, but the sum of the three angles of a spherical triangle is not constant.

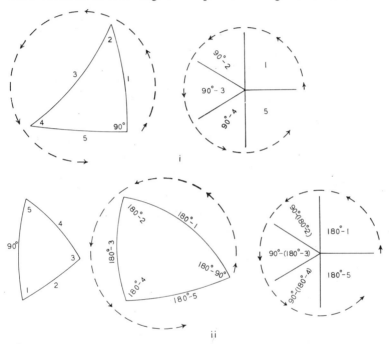

FIG. 89 Napier's device for the solution of right-angled spherical triangles.

Solution of right-angled triangles

Solution is simplified if one part, whether an angle or a side of the three known parts = 90°. The frequent use which we have made of orthogonal relationships, notably that of a great circle and its pole, means that we can work many of the simpler problems exclusively in terms of such right-angled triangles, and we shall first describe the appropriate procedure and illustrate its application to some examples (cf. Phillips, 1963, pp. 183–4, from which the following paragraphs are quoted by permission of the publishers).

It is unnecessary to remember a number of distinct formulae appropriate to the various possible cases, for Napier long ago described a device by which the required formula can be written down at sight. Since this device applies only to right-angled triangles, such triangles are often termed *Napierian triangles*. In Fig. 89.i the five parts of the triangle (excluding the right angle) are numbered in order 1–5, as they are encountered by moving around the triangle from the right angle. On the right of the figure is sketched a five-compartment diagram, three compartments on the left of the vertical stroke each filled in with the symbol $90°-$, and two on the right. Each time a Napierian triangle is to be solved, this device is sketched; the numbered parts of the triangle are then written in the appropriate compartments, starting from the horizontal stroke as we start from the right-angled part of the triangle and proceeding from compartment to compartment as we proceed around the triangle. Fig. 89.ii shows the slight modification necessary for the case in which the $90°$ element is a side of the original triangle—the Napierian device is filled in terms of the polar triangle, the sides and angles of which are respectively the supplements of the corresponding angles and sides of the original triangle. Since the triangle is soluble, two of the parts 1, 2, 3, 4 and 5 are already of known magnitude; any other part which we may require must be situated *either* in such a way that it and the two known parts are all three in adjacent compartments *or* in such a way that two of the compartments in question are opposite the third. All the required formulae are then summarized in the statements:—

The sine of a middle part = { The product of the tangents of adjacent parts
 or
 The product of the cosines of opposite parts.

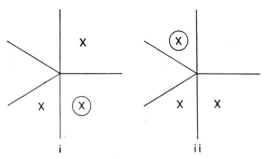

Fig. 90   Diagram of the arrangement of parts in Napier's device.

In Fig. 90.i, the arrangement is that of a middle part (which is ringed) with two adjacent parts, whilst Fig. 90.ii shows a middle part with two opposites. (Note that it is not necessary that the unknown

CALCULATIONS BY SPHERICAL TRIGONOMETRY

should be the middle part of the formula; hence an intermediate auxiliary solution is never necessary—we can always go straight to the required answer for any of the three unknown parts.)

To illustrate the use of these methods we may look back at our first problem, the determination of apparent dips of the bed in Fig. 11, p. 8. The strike of the bed is 300° and angle of dip 30°. Hence to calculate the arc ab we use the triangle:—

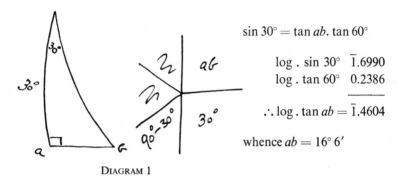

DIAGRAM 1

$\sin 30° = \tan ab . \tan 60°$

$\log . \sin 30°\quad \bar{1}.6990$
$\log . \tan 60°\quad 0.2386$

$\therefore \log . \tan ab = \bar{1}.4604$

whence $ab = 16° 6'$

Similarly, for cd:—

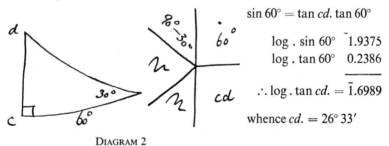

DIAGRAM 2

$\sin 60° = \tan cd . \tan 60°$

$\log . \sin 60°\quad \bar{1}.9375$
$\log . \tan 60°\quad 0.2386$

$\therefore \log . \tan cd. = \bar{1}.6989$

whence $cd. = 26° 33'$

and for ef:—

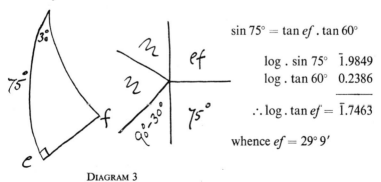

DIAGRAM 3

$\sin 75° = \tan ef . \tan 60°$

$\log . \sin 75°\quad \bar{1}.9849$
$\log . \tan 60°\quad 0.2386$

$\therefore \log . \tan ef = \bar{1}.7463$

whence $ef = 29° 9'$

The relationship of pitch to plunge (Fig. 14, p. 11) similarly involves Napierian solutions:—

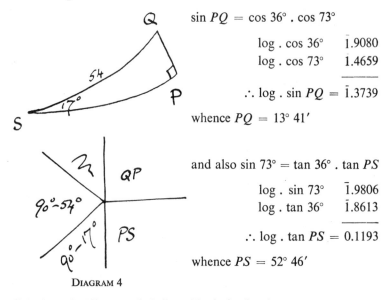

DIAGRAM 4

sin PQ = cos 36° . cos 73°

log . cos 36°	$\bar{1}.9080$
log . cos 73°	$\bar{1}.4659$
∴ log . sin PQ =	$\bar{1}.3739$

whence PQ = 13° 41′

and also sin 73° = tan 36° . tan PS

log . sin 73°	$\bar{1}.9806$
log . tan 36°	$\bar{1}.8613$
∴ log . tan PS =	0.1193

whence PS = 52° 46′

Solution of oblique-angled ('non-Napierian') triangles

Though simple problems can be worked exclusively in terms of right-angled triangles, we shall frequently encounter also data so related that at least at the outset of the calculations we must solve an oblique-angled triangle.

Frequently we shall be given two sides a and b and the included angle C (as with two apparent dips of known azimuths, or the azimuths and plunges of two oblique boreholes). To calculate the third side c we use the formula:—

$$\cos c = \cos a . \cos b + \sin a . \sin b . \cos C \qquad \text{i}$$

If the three sides a, b and c are given, a situation which also arises in many borehole problems, we can calculate an angle more conveniently from the formula*:—

$$\tan \frac{A}{2} = \sqrt{\frac{\sin(s-b).\sin(s-c)}{\sin s . \sin(s-a)}} \qquad \text{ii}$$

where $s = \dfrac{a+b+c}{2}$

*If the reader should be familiar with the methods and tables used in navigation he may prefer to use the formula:—

$$\text{hav } A = \frac{\text{hav } a - \text{hav}(b\sim c)}{\sin b . \sin c}$$

CALCULATIONS BY SPHERICAL TRIGONOMETRY

A further useful relationship is:—

$$\frac{\sin a}{\sin A} = \frac{\sin b}{\sin B} = \frac{\sin c}{\sin C} \qquad \text{iii}$$

Finally, given two angles A and B and the included side c, and being required to find the sides a and b, it is convenient to work with the formulae:—

$$\tan\frac{a+b}{2} = \frac{\cos\frac{A-B}{2}}{\cos\frac{A+B}{2}} \tan\frac{c}{2}$$

$$\tan\frac{a-b}{2} = \frac{\sin\frac{A-B}{2}}{\sin\frac{A+B}{2}} \tan\frac{c}{2} \qquad \text{iv}$$

To illustrate the application of some of these formulae we proceed to calculate the orientation of the normal OP^1 in the borehole problem, Fig. 63, p. 46.

Labelling the centre of the stereogram O, we first solve the triangle OHJ by an application of formula i above:—

$$\cos HJ = \cos 30° . \cos 45° + \sin 30° . \sin 45° . \cos 110°$$

$$= 0.6124 - 0.1210 \qquad \log . \cos 30° \quad \bar{1}.9375$$
$$\log . \cos 45° \quad \bar{1}.8495$$

$$\overline{\bar{1}.7870}$$

$$= 0.4914$$

$$\log . \sin 30° \quad \bar{1}.6990$$
$$\log . \sin 45° \quad \bar{1}.8495$$
whence $HJ = 60° \ 34'$ $\qquad \log. \cos 110° \quad \bar{1}.5341$

$$\overline{\bar{1}.0826}$$

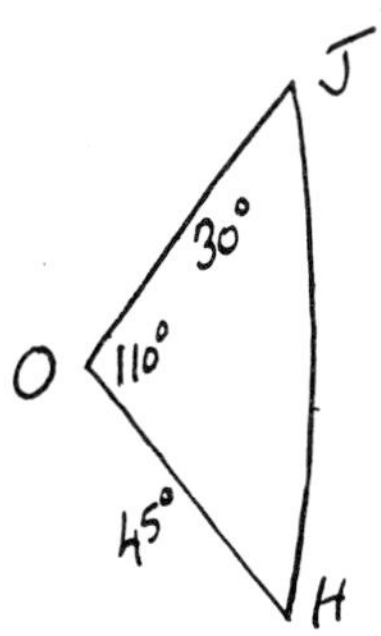

DIAGRAM 5

and from formulae iii,

$$\sin \angle OHJ = \frac{\sin 30° . \sin 110°}{\sin 60° \ 34'}$$

$$\log . \sin 30° \qquad \bar{1}.6990$$
$$\log . \sin 110° \qquad \bar{1}.9730$$
$$\log . \operatorname{cosec} 60° \ 34' \quad 0.0600$$

whence $\angle OHJ = 32° \ 39'$ $\qquad \therefore \log . \sin OHJ = \bar{1}.7320$

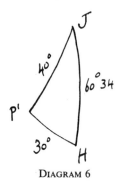

DIAGRAM 6

We next solve the triangle P^1HJ by application of formula ii:—

$$HP^1 = 30°$$
$$JP^1 = 40°$$
$$HJ = 60° \ 34'$$
$$2 \overline{)130° \ 34'}$$
$$s = 65° \ 17'$$

$$\text{Tan} \frac{\angle P^1HJ}{2} = \sqrt{\frac{\sin 35° \ 17' \cdot \sin 4° \ 43'}{\sin 65° \ 17' \cdot \sin 25° \ 17'}}$$

log . sin 35° 17'	$\bar{1}.7616$
log . sin 4° 43'	$\bar{2}.9151$
log . cosec 65° 17'	0.0417
log . cosec 25° 17'	0.3695
	$2\overline{)\bar{1}.0879}$

$$\therefore \log . \tan \frac{P^1HJ}{2} = \bar{1}.5439$$

whence $\dfrac{\angle P^1HJ}{2} = 19° \ 17'$

and $\angle P^1HJ = 38° \ 34'$

These solutions now allow us to solve the triangle OHP^1 for OP^1 (giving the required plunge) and $\angle P^1OH$ (giving the required azimuth):—

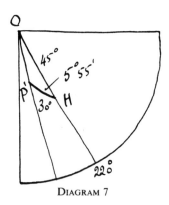

DIAGRAM 7

$\cos OP^1 = \cos 30° \cdot \cos 45° + \sin 30° \cdot \sin 45° \cdot \cos 5° \ 55'$

$\qquad = 0.6124 + 0.3518 \qquad \bar{1}.6990$
$\qquad\qquad\qquad\qquad\qquad\qquad\qquad \bar{1}.8495$
$\qquad = 0.9642 \qquad\qquad\qquad\qquad \bar{1}.9977$

whence $OP^1 = 15° \ 23'$ (i.e., the normal plunges 74° 37′ from the horizontal.)

$\qquad\qquad\qquad\qquad\qquad\qquad\qquad \bar{1}.5462$

and also:—

$$\sin \angle P^1OH = \frac{\sin 30° \cdot \sin 5° 55'}{\sin 15° 23'}$$

$\bar{1}.6990$
$\bar{1}.0133$
0.5763
———
$\bar{1}.2886$

whence $\angle P^1OH = 11° 12'$

The azimuth of plunge of the hole H being 220°, that of the normal OP^1 is hence 231°.

Finally, as an illustration of the use of formulae iv, we may calculate the plunge of the axis of the fold in Fig. 19, p. 14:—

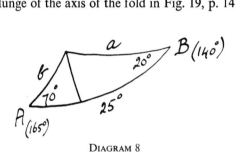

DIAGRAM 8

$$\tan \frac{a+b}{2} = \frac{\cos 25° \cdot \tan 12° 30'}{\cos 45°}$$

$\bar{1}.9573$
$\bar{1}.3458$
0.1505
———
$\bar{1}.4536$

whence $\frac{a+b}{2} = 15° 52'$

$$\tan \frac{a-b}{2} = \frac{\sin 25° \cdot \tan 12° 30'}{\sin 45°}$$

$\bar{1}.6259$
$\bar{1}.3458$
0.1505
———
$\bar{1}.1222$

whence $\frac{a-b}{2} = 7° 33'$

and hence $a = 23° 25'$

We are now in a position to determine the required azimuth and plunge from two Napierian solutions:—

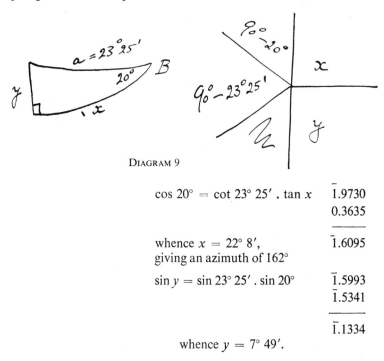

DIAGRAM 9

$$\cos 20° = \cot 23° 25' \cdot \tan x \qquad \bar{1}.9730$$
$$0.3635$$

$$\overline{}$$

whence $x = 22° 8'$, $\qquad \bar{1}.6095$
giving an azimuth of 162°

$\sin y = \sin 23° 25' \cdot \sin 20° \qquad \bar{1}.5993$
$\bar{1}.5341$

$$\overline{}$$

$\bar{1}.1334$

whence $y = 7° 49'$.

(Morishita, 1938, p. 232, works a further example of this method in detail in the solution of a fault-problem.)

BIBLIOGRAPHY

(Entries in **heavy type** indicate that the work is specifically concerned with the stereographic method.)

Anderson, E. M. 1951. *The Dynamics of faulting.* 2nd. edition, Oliver and Boyd, Edinburgh and London.

Addie, G. 1962. **A simplified stereographic projection solution of the two drillhole problem.** *Econ. Geol.,* vol. 57, pp. 1122–6.

Addie, G. 1964. **A simple solution of the two drillhole problem giving one answer.** *Econ. Geol.,* vol. 59, pp. 149–53.

Badgley, P. C. 1959. *Structural methods for the exploration geologist.* (Chap. 8, pp. 187–242. **The use of stereographic projections in solving structural problems.**) Harper Bros., New York.

Beckwith, R. H. 1947. **Fault problems in fault planes.** *Bull. Geol. Soc. Amer.,* vol. 58, pp. 79–108.

Billings, M. P. 1942 (2nd edition 1954). *Structural geology.* Prentice-Hall, New York.

Boucher, F. G., Hildebrandt, A. B. and Hagen, H. B. 1950. New dip logging method. *Bull. Amer. Assoc. Petrol. Geol.,* vol. 34, pp. 2007–26.

Brett, G. W. 1955. Cross-bedding in the Baraboo quartzite of Wisconsin. *Journ. Geol.,* vol. 63, pp. 143–8.

Bucher, W. H. 1920. The mechanical interpretation of joints. Part I. *Journ. Geol.,* vol. 28, pp. 707–30.

Bucher, W. H. 1943. **Dip and strike from three not parallel drill cores lacking key beds.** *Econ. Geol.,* vol. 38, pp. 648–57.

Bucher, W. H. 1944. **The stereographic projection, a handy tool for the practical geologist.** *Journ. Geol.,* vol. 52, pp. 191–212.

Challinor, J. A. 1944. A note on tilted folds and direction of pitch. *Proc. Geol. Assoc.,* vol. 55, pp. 94–8.

Clark, R. H. and McIntyre, D. B. 1951. The use of the terms Pitch and Plunge. *Amer. Journ. Sci.,* vol. 249, pp. 591–9.

Clark, R. H. and McIntyre, D. B. 1951a. **A macroscopic method of fabric analysis.** *Amer. Journ. Sci.,* vol. 249, pp. 755–68.

Cotton, L. A. and Garretty, M. D. 1945. **Use of the stereographic projection in solving problems in structural and mining geology.** *Geological Bulletin* No. 2, North Broken Hill Ltd. (Roneoed.)

Dahlstrom, C. D. A. 1954. Statistical analysis of cylindrical folds. *Canad. Mining Metall. Bull. Trans.,* vol. 57, pp. 140–5.

De Chambrier, P. 1953. The Microlog continuous dipmeter. *Geophysics,* vol. 18, pp. 929–50.

De Cizancourt, H. 1947. Quelques problèmes de tectonique géométrique. Deuxième partie: Les fractures des roches. *Rev. de l'Inst. Franc. du Petrole,* vol. II, n°. 2, pp. 81–98, n°. 3, pp. 141–50.

Den Tex. E. 1953. **Stereographic distinction of linear and planar structures from apparent lineations in random exposure planes.** *Journ. Geol. Soc. Australia,* vol. 1, pp. 55–66.

De Witte, A. J. 1956. A graphical method of dipmeter interpretation using the stereonet. *Journ. Petrol. Tech.*, vol. 8, no. 8.

Donn, W. L. and Shimer, J. A. 1958. *Graphic methods in structural geology.* Appleton-Century-Crofts Inc., New York.

Eardley, A. T. 1938. Graphic treatment of folds in three dimensions. *Bull. Amer. Assoc. Petrol. Geol.*, vol. 22, pp. 483–9.

Earle, K. W. 1934. *Dip and strike problems mathematically surveyed.* Murby and Co., London.

Fischer, G. 1930. Statistische Darstellungsmethoden in der tektonischen Forschung. *Sitzungsber. (Preuss.) Geol. Landesanstalt*, Heft 5, pp. 4–25.

Fisher, D. J. 1938. **Problem of two tilts and the stereographic projection.** *Bull. Amer. Assoc. Petrol. Geol.*, vol. 22, pp. 1261–71

Fisher, D. J. 1941. **Drillhole problems in the stereographic projection.** *Econ. Geol.*, vol. 36, pp. 551–60.

Fisher, D. J. 1941a. **A new projection protractor.** *Journ. Geol.*, vol. 49, pp. 292–323, 419–42.

Fisher, D. J. 1943. **Measuring linear structures on steep-dipping surfaces.** *Amer. Miner.*, vol. 28, pp. 204–8.

Fisher, D. J. 1952. Crystallographic projections nomenclature dilemma. *Amer. Miner.*, vol. 37, pp. 857–61.

Fisher, D. J. 1958. **Graphical determination of the dip in deformed and cleaved rocks; a discussion.** *Journ. Geol.*, vol. 66, p. 100.

Gilluly, J. 1944. **Dip and strike from three not parallel drill cores lacking key beds.** *Econ. Geol.*, vol. 39, pp. 359–63.

Hamann, P. J. 1958. **Bestimmung der Bewegungsrichtung an Verwerfungen mit Hilfe des Schmidt'schen Netzes.** *Notizbl. hess. L.-Amt. Bodenforsch.*, Band 87, pp. 188–91.

Haman, P. J. 1961. **Manual of the stereographic projection for a geometric and kinematic analysis of folds and faults.** *West Canadian Res. Pub. of Geology and related Sciences.* Ser. 1, no. 1.

Harrison, P. W. 1957. **New technique for three-dimensional fabric analysis of till and englacial debris . . .** *Journ. Geol.*, vol. 65, pp. 98–105.

Higgs, D. V. and Tunell, G. 1959 (2nd edition 1966). **Angular relations of lines and planes.** W. H. Freeman and Co., San Francisco and London.

Ho Tso-lin. 1957. **Determination of rotational fault plane.** *Scientia Sinica*, vol. 6, pp. 169–77.

Hobson, G. D. 1943. A graphical solution of the problem of two tilts. *Proc. Geol. Assoc.*, vol. 54, pp. 29–32.

Hobson, G. D. 1945. The application of tilt to a fold. *Proc. Geol. Assoc.*, vol. 55 (for 1944), pp. 216–21.

Hubbert, M. K. 1931. Graphic solution of strike and dip from two angular components. *Bull. Amer. Assoc. Petrol. Geol.*, vol. 15, pp. 283–6.

Ingerson, E. 1942. **Apparatus for direct measurement of linear structures.** *Amer. Miner.*, vol. 27, pp. 721–5.

Ingerson, E. 1944. **Why petrofabrics?** *Trans. Amer. Geophys. Union*, pp. 635–52.

Ives, R. L. 1939. Measurements in block diagrams. *Econ. Geol.*, vol. 34, pp. 561–572.

BIBLIOGRAPHY

Johnson, C. H. 1939. **New mathematical and 'stereographic net' solutions to problems of two tilts—with applications to core orientation.** *Bull. Amer. Assoc. Petrol. Geol.*, vol. 23, pp. 663–85.

Johnston, W. D., jun. and Nolan, T. B. 1937. Isometric block diagrams in mining geology. *Econ. Geol.*, vol. 32, pp. 550–69.

Krumbein, W. C. 1939. Preferred orientation of pebbles in sedimentary deposits. *Journ. Geol.*, vol. 47, pp. 673–706.

Lobeck, A. K. 1924. *Block Diagrams.* John Wiley & Sons, New York. (1958. Emerson-Trussell, Amherst.)

Lowe, K. E. 1946. **A graphic solution for certain problems of linear structure.** *Amer. Miner.*, vol. 31, pp. 425–34.

McClellan, H. 1948. **Core orientation by graphical and mathematical methods** (with a Discussion by C. H. Johnson). *Bull. Amer. Assoc. Petrol. Geol.*, vol. 32, pp. 262–82.

McLaughlin, K. P. 1948. Secondary tilt: a review and a new solution. *Journ. Geol.*, vol. 56, pp. 72–4.

Mead, W. J. 1921. Determination of attitude of concealed bedded formations by diamond drilling. *Econ. Geol.*, vol. 16, pp. 37–47.

Mills, J. W. 1963. **A simplified stereographic projection solution of the two drillhole problem.** *Econ. Geol.*, vol. 58, pp. 618–21.

Morishita, M. 1938. **On the graphic method of representing faults and strata.** *Jap. Journ. Geol. Geogr.*, vol. 15, pp. 207–39.

Müller, L. 1933. Untersuchungen über statistische Kluftmessung. *Geologie u. Bauwesen*, Jahrg. 5, pp. 185–255.

Nevin, C. M. 1949. *Principles of Structural Geology.* 4th edition. John Wiley & Sons, New York.

Nobel, D. C. and Eberly, S. W. 1964. A digital computer procedure for preparing beta diagrams. *Amer. Journ. Sci.*, vol. 262, pp. 1124–9.

Pettijohn, F. J. 1957. Palaeocurrents of Lake Superior Precambrian quartzites. *Bull. Geol. Soc. Amer.*, vol. 68, pp. 469–80.

Phillips, F. C. 1963. *An Introduction to crystallography.* Longmans, London.

Price, N. J. 1966. *Fault and joint development in brittle and semi-brittle rock.* Pergamon Press, Oxford.

Ragan, D. M. 1968. *Structural geology. An introduction to geometrical techniques.* John Wiley & Sons, New York

Ramsay, J. G. 1960. **The deformation of early linear structures in cases of repeated folding.** *Journ. Geol.*, vol. 68, pp. 75–93.

Ramsay, J. G. 1961. **The effects of folding upon the orientation of sedimentation structures.** *Journ. Geol.*, vol. 69, pp. 84–100.

Ramsay, J. G. 1962. **The geometry of conjugate fold systems.** *Geol. Mag.*, vol. 99, pp. 516–26.

Ramsay, J. G. 1964. The uses and limitations of beta-diagrams and pi-diagrams in the geometrical analysis of folds. *Quart. Journ. Geol. Soc.*, vol. 120, pp. 435–54.

Ramsay, J. G. 1967. *Folding and fracturing of rocks.* McGraw Hill, New York and London.

Robinson, P. et al. 1963. Preparation of beta-diagrams in structural geology by a digital computer. *Amer. Journ. Sci.*, vol. 261, pp. 913–28.

Rodgers, J. 1952. Use of equal-area and other projections in the statistical treatment of joints. *Bull. Geol. Soc. Amer.*, vol. 63, pp. 427–30.

Sander, B. 1930. *Gefügekunde der Gesteine*. Vienna.

Sander, B. 1942. Über Flächen- und Achsengefüge. *Mitt. Reichs. Bodenf. Zweigst. Wien.*, 4, pp. 1–94.

Sander, B. 1948. *Einführung in die Gefügekunde der geologischen Korper*. I Teil. Springer Verlag, Vienna and Innsbruck. (English translation, see Sander, 1970.)

Sander, B. 1950. *Einführung in die Gefügekunde der geologischen Korper*. II Teil. Springer Verlag, Vienna and Innsbruck. (English translation, see Sander, 1970.)

Sander, B. 1970. *An Introduction to the Study of Fabrics of Geological Bodies*. (Authorized translation from the German of parts I and II, 1948 and 1950, by F. C. Phillips and G. Windsor.) Pergamon Press, Oxford.

Seitz, O. 1924. **Das Wulffsche Netz als Hilfsmittel bei tektonischen Untersuchungen.** *Berg. -und Hütt. Zeitschrift 'Glückauf'*, Nr. 19.

Sokol, R. 1927. *Geologisches Praktikum*, Berlin.

Stach, E. 1923. Die stereographische Darstellung tektonischer Formen im 'Würfeldiagramm' auf 'Stereo-Millimeterpapier'. *Zeit. d. Deut. Geol. Ges.*, Bd. 74 (for 1922), pp. 277–309.

Stach, E. 1929. Die Zeichnung geologischer und bergbaulicher Raumbilder. *Berg. -und Hütt. Zeitschrift. 'Glückauf'*.

Stein, H. A. 1941. A trigonometric solution of the two-drillhole problem. *Econ. Geol.*, vol. 36, 84–94.

Stockwell, C. H. 1950. **The use of plunge in the construction of cross-sections of folds.** *Proc. Geol. Assoc. Canada*, vol. 3, pp. 97–121.

Terpstra, P. 1948. (**Some examples of the use of a stereonet—English abstract.**) *Geol. en Mijnbouw*, jg. 10, no. 11, pp. 303–7.

Turner, F. J. and Weiss, L. E. 1963. *Structural analysis of metamorphic tectonites*. McGraw-Hill, New York and London.

Wager, L. R. and Deer, W. A. 1934. The petrology of the Skaergaard intrusion, Kangerdlugssuaq, East Greenland. *Medd. Grønland*, Bd. 105, Nr. 4.

Wallace, R. E. 1948. **A stereographic calculator.** *Journ. Geol.*, vol. 56, 488–90.

Wallace, R. E. 1950. **Determination of dip and strike by indirect observations in the field and from aerial photographs.** *Journ. Geol.*, vol. 58, pp. 269–80.

Wallace, R. E. 1951. Geometry of shearing stress and relation to faulting. *Journ. Geol.*, vol. 59, pp. 118–30.

Wegmann, C. E. 1929. Über alpine Tektonik und ihre Anwendung auf das Grundgebirge Finnlands. *Bull. Comm. géol. Finl.*, no. 85, pp. 49–53.

Wegmann, C. E. 1929a. Stereogramm des Gebietes von Soanlahti-Suistamo. *Bull. Comm. géol. Finl.*, no. 85, pp. 58–66.

Wegmann, C. E. 1929b. Beispiele tektonischer Analysen des Grundgebirges in Finnland. *Bull. Comm. géol. Finl.*, no. 87, pp. 98–127.

Weiss, L. E. 1959. **Geometry of superposed folding.** *Bull. Geol. Soc. Amer.*, vol. 70, pp. 91–106.

Whitten, E. H. T. 1957. **A note on stereographic analysis of bedding planes with special reference to folding in the Isle of Wight, England.** *Journ. Geol.*, vol. 65, pp. 551–6.

Whitten, E. H. T. 1966. *Structural geology of folded rocks*. Rand McNally, Chicago.

BIBLIOGRAPHY

Williams, A. 1958. **Oblique-slip faults and rotated stress systems.** *Geol. Mag.*, vol. 95, pp. 207–18.

Wisser, E. 1932. An aid in the interpretation of diamond drill cores. *Econ. Geol.*, vol. 27, pp. 437–49.

Wollnough, W. G. and Benson, W. N. 1957. Graphical determination of the dip in deformed and cleaved sedimentary rocks. *Journ. Geol.*, vol. 65, pp. 428–33.

Zimmer, P. W. 1963. Orientation of small diameter drill core. *Econ. Geol.*, vol. 58, pp. 1313–25.

EXERCISES

(Answers on p. 86)

1. A bedding-plane dips 28° on a bearing 140°. What is the apparent dip (*a*) to the south (*b*) on a bearing 212°?

2. Two apparent dips of a structural plane are measured as 16° on a bearing 120° and 22° on a bearing 227°. What is the apparent dip due east?

3. A marker horizon in a succession of dipping plane strata is encountered at a depth of 208 ft. above O.D. in a vertical bore. In a second bore, also vertical and located 500 ft. south-west of the first, the marker is 102 ft. above O.D., whilst in a third bore due south of the first and distant 300 ft. from it the level is 49 ft. above O.D. Determine the strike and true dip of the strata.

4. In a lode dipping 60° on a bearing 305° an ore-shoot pitches 35° south-westerly. Determine the bearing, and angle, of plunge of the ore-shoot.

5. An ore-shoot is developed at the intersection of a quartz vein dipping 28° on a bearing 292° with a fault striking 80° and hading 25° southerly. Determine the bearing and plunge of the ore-shoot and its angles of pitch on the plane of the vein and on the fault-plane.

6. On one limb of a plunging anticline the strike is 35° and the dip 24° south-easterly, whilst on the opposite limb the strike is 76° and the dip 34° northerly. Find the bearing and angle of plunge of the crest of the fold.

7. A sedimentary succession, in which the general strike is W–E but the bedding is obscure, is cut by a fault hading 30° on a bearing 210°. Slickensiding on the fault-plane pitches 80° south-westerly. If a pebbly layer within the sediments appears to cross the outcrop of the fault without offset what is the dip of the strata?

8. In a deeply dissected terrain the outcrop of a marker horizon on a bedded series is observed from a distance to make a marked 'V' across a valley falling south-westerly. Viewed from a point on lower ground due south the lines of sight slope upwards at an elevation of 10° above horizontal: the apparent dip of the western line of outcrop is read as 15° easterly, that of the eastern portion of the outcrop as 20° westerly. From a second station, on a peak from which the bearing of the exposure is 20°, the lines of sight are depressed 15° below horizontal, and the apparent dips of the two lines of outcrop are read as 34° easterly and 46° westerly respectively. Determine the strike and true dip of the marker horizon, assuming that it belongs to a plane dipping series.

9. On various joint faces of an igneous rock in which a linear structure is present the following observations of lineations were made:—

	Strike of joint	Dip	Pitch of lineation
1.	146°	20° SW	35° NW
2.	12°	62° E	12° S
3.	85°	40° N	38° NW
4.	97°	65° S	17° W
5.	33°	64° SE	No clear lineation observed.
6.	75°	15° N	46° NW

   Determine the orientation of the linear structure and comment on the result.

10. On the foreset beds of a false-bedded series now dipping 20° on a bearing 244° the dip is measured as 34° on a bearing 132°. What was the original direction and amount of dip of the foreset beds at the time when the main bedding-planes were horizontal?

EXERCISES

11. On opposite limbs of a fold, developed beneath a plane of unconformity now dipping 22° on a bearing 220°, the following observations were made: eastern limb dip 23° on a bearing 132°, western limb dip 46° on a bearing 250°. Investigate the nature of the original folding, before the plane of unconformity was tilted.

12. On one limb of an anticline, dipping 22° on a bearing 210°, the pitch of a lineation is 50° westerly; on the opposite limb, dipping 40° on a bearing 353°, there is an almost horizontal lineation. Could this be interpreted as a previously existing lineation on a plane involved in later folding?

13. In an area in which the regional strike of a concealed structure is known to be N–S the following data were obtained from two inclined boreholes:—

	Plunge	Bearing	Angle of bedding-normal and core-axis
Borehole A	48°	210°	60°
Borehole B	34°	130°	35°

Determine the strike and amount of dip of the strata at this location.

14. The following data were obtained from three non-parallel boreholes put down to intersect a bedded series in depth:—

	Plunge	Bearing	Angle of bedding-normal and core-axis
Borehole A	25°	235°	42°
Borehole B	40°	150°	70°
Borehole C	24°	50°	30°

Determine the strike and dip of the strata.

15. On the eastern side of a fault, striking 145° and hading 38° SW, a symmetrical syncline is exposed; the strike of the horizontal axis is 10° and the dips on the limbs 50°. On the western (downthrow) side of the fault this structure appears as an asymmetric plunging fold, of which the western limb strikes 157° and dips 42° NE. Investigate the nature of the movement on the fault-plane and determine the plunge of the axis of the syncline in the downthrown block.

16. In a trial shaft sunk through interbedded slates and uncleaved grits the bedding-planes are found to dip 52° on a bearing 227°, whilst cleavage in the slates dips 25° on a bearing 277°. What is the probable nature of the folding in this region?

17. In a region of folded strata the fold axes plunge 12° on a bearing 250°. The rocks are cut by a set of shear planes dipping 50° SW. In a core drawn from a borehole plunging 40° N the trace of bedding can be seen on the elliptical cross-section where the core is broken by a shear plane, and pitches S.S.Easterly at an angle of 35° from the southerly extremity of the minor axis of the ellipse. What is the dip of the strata at this point in the borehole?

18. In an area of steeply-folded strata the axes plunge NW at 15°. From boreholes put down at three different locations through a cover of superficial deposits the following observations are made:—

	Plunge	Bearing	Angle of bedding-normal and core-axis
Borehole A	50°	260°	50°
Borehole B	46°	29°	35°
Borehole C	Vertical bore		70°

Determine the directions and amounts of dip at these three places.

Two intersecting shear-zones have the orientations—strike 101°, dip 60° N and strike 40°, dip 30° SE respectively. Determine the bearing and plunge of a trial hole which would cut their line of intersection perpendicularly, passing (for safety) midway between the zones until the line of intersection is encountered.

Answers to Exercises

1. (a) 22°; (b) 10°.
2. 0° (the plane strikes W–E).
3. Strike 66°, dip 30° south-easterly.
4. Plunge 30° on a bearing 234°.
5. Ore-shoot plunges 21° on a bearing 250°; pitch on vein 51° southerly, on fault-plane 24° westerly.
6. Plunge 11° on a bearing 60°.
7. 68° S. (Since the bed is not offset in plan movement on the fault must have been parallel to the line of intersection of the fault with the bedding.)
8. Strike 121°, dip 24° south-westerly.
9. Linear structure plunges 25° on a bearing 295°. Plane 5 is almost at right angles to this structure.
10. Dip 45° on a bearing 111°.
11. Symmetrical folding about horizontal N–S axes.
12. Yes.
13. Dip 30° on a bearing 277°.
14. Strike 117°, dip 83° south-westerly, or strike 166°, dip 79° westerly. The three holes are badly chosen; A and B, for example, give four possible results and if these had been worked out before C had been run it should have been noticed that the orientation selected for C is so nearly symmetrical to two of the results from A and B that it cannot discriminate between them.
15. Rotation during slip of 25° anticlockwise about the normal to the fault-plane as viewed from the west. Plunge of the axis 16° on a bearing 355°.
16. Plunging folds with reversed limbs. The geometrical axis plunges 24° on a bearing 296°.
17. 30° S.
18. Draw the great circle of which the fold-axis is the normal; the poles of normals to beds must lie on this circle in every part of the area.
    At A the dip is 79° on a bearing 41°.
    At B the dip is 77° on a bearing 228°.
    At C the dip is 70° on a bearing 50° or 219° (the choice of a vertical borehole leaves an ambiguity).
19. Plunge 61° on a bearing 227°.

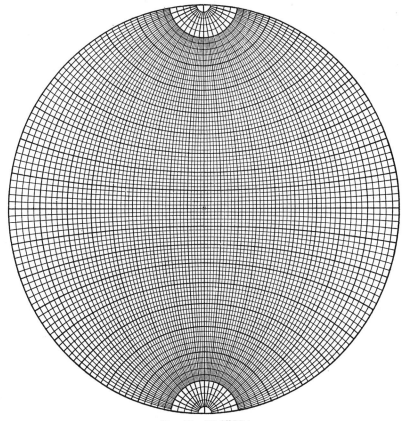

FIG. 91 Wulff Net.

INDEX

Accuracy of graphical solution, 69
ADDIE, G., 42
ANDERSON, E. M., 25
Answers to exercises, 86
Apparent dip, 7
   on intersecting planes, 13
Auxiliary rotation, 36
Auxiliary solutions, 54

B-axes, 66, 68
BADGLEY, P. C., iv, 27, 53, 54, 60
BENSON, W. N., 7
BECKWITH, R. H., 54, 56
β-diagram, 68
Bibliography, 79
BILLINGS, M. P., iv, 9, 54, 60, 64
Block-diagrams, 56
Boreholes, 30, 42
BOUCHER, F. G., 51
BRETT, G. W., 30
BUCHER, W. H., iii, 9, 12, 25, 29, 42

Cabinet projection, 58
Calculations by spherical trigonometry, 70
CHALLINOR, J. A., 31
CLARK, R. H., 9, 12, 23
Clinographic projection, 58
Core-bedding angle, 43
Corresponding points, 54
COTTON, L. A., iv, 27
Counting out a pole-diagram, 64
Cyclographic projection, 17

DE CIZANCOURT, H., 60, 61
DEER, W. A., 30
DEN TEX, E., 12
Dipmeter, 9, 31, 51
Directions of principal stress, 25
Distant outcrop, 8, 15, 17
DONN, W. L., iv, 7
Drill cores, 30, 42

EARLE, K. W., 7, 9, 12, 29
EBERLEY, S. W., 69
Equal-area projection, 60, 61
Exercises, 84

False-bedding, 30
Fault-planes, calculation in, 54

Fault-systems, 23
Faults, 39, 54
Federov net, 63
FISCHER, G., 64, 65
FISHER, D. J., iv, 9, 12, 17, 29, 42
Fold-systems, 66

GARRETTY, M. D., iv, 27
GILLULY, J., 42
Great circle, 1

Haarmann angles, 54
Hade of a fault, 15
HAGAN, H. B., 51
HAMAN, P. J., 54, 56
HARRISON, P. W., 62
HIGGS, D. V., 70
HILDEBRANDT, A. B., 51
HO, TSO-LIN., 40
HOBSON, G. D., 29, 31
HUBBERT, M. K., 7

Inclined boreholes, 42
INGERSON, E., 9, 10, 12
Interpretation of cores, 42
Intersecting planes, 12
Isometric projection, 56
IVES, R. L., 56

JOHNSON, C. H., 29, 31
JOHNSTON, W. D., 56
Joint-systems, 23, 64

KLEEMAN, A. W., iv
Kohlschütter apparatus, 69
KRUMBEIN, W. C., 62

Lambert projection, 60
Linear structure, 20, 23
Lineation, 9, 20
   on folds, 32, 40, 66
   on oblique planes of section, 20
   on planes, 9
   trend, 10
LUBECK, A. K., 56, 58
LOWE, K. E., 12, 21

MCINTYRE, D. B., 9, 12, 23
MCCLELLAN, H., 31

INDEX

McLaughlin, K. P., 29
Mead, W. J., 42
Meridional stereographic net, 5
Mills, J. W., 42, 48
Moine schists, 67
Morishita, M., 54, 78
Müller, L., 60, 64

Napierian triangles, 72
Net slip, 54
Nevin, C. M., iv, 7, 12, 25
Nobel, D. C., 69
Nolan, T. B., 56
Non-parallel boreholes, 42
N-plane, 21

Oblique aerial photographs, 17
Oblique-angled triangles, 74
Oblique boreholes, 42

Pettijohn, F. J., 30, 35
Phillips, F. C., 58, 71
Pitch, 9
    angle of, 10
Pitching fold—see plunging fold
Planar structure, 20, 23
Plunge, 9
    angle of, 10
    of fold axis, 66
Plunging fold, 13, 31, 66
Polar nets, 61, 62
Polar projection, 17, 30
Pole-diagram, 64
Pole of a plane, 17
π-pole, 68
Price, N. J., 27
Primitive, 2
Principal diameters, 7
Problem of two tilts, 29

Ragan, D. M., iv, 54
Rake, 10
Ramsay, J. G., 27, 35, 60, 69
Resultant path, 54
Robinson, P., 69
Rodding, 20
Rodgers, J., 60
Rotation of sphere, 28, 36

Sander, B., iv, 9, 60, 68
Scatter-diagram, 64
Schmidegg counter, 65
Schmidt net, 60, 61
Scissor faulting, 39, 55
Secondary tilt, 29
    of a fold, 31
Seitz, O., iii, 60
Shimer, J. A., iv, 7
Small circle, 5
Sokol, R., iii, 7, 54
Spherical projection, 1
Spherical trigonometry, 70
s-planes, 68
Stach, E., 56
Start Point, 64, 65, 66
Stein, H. A., 42
Stereogram, 2, 17
Stereographic net, 5, 6, 60, 69, 87
Stereographic projection, principle, 1
Stockwell, C. H., 7, 69
Strain ellipsoid, 23

Tectonic syntheses, 60
Terpstra, P., 29, 31
Three-point problem, 8
Tilt of a fold, 31
Trend of lineation, 10
True dip from apparent dips, 7, 13, 52
Tunell, G., 70
Turner, F. J., 51, 60, 68

Upper hemisphere, projection, 33

Veins, intersecting, 12

Wager, L. R., 30
Wallace, R. E., 12, 17, 27, 56, 69
Wegmann, C. E., iii, 56, 60, 66, 67
Weiss, L. E., 51, 60, 68
Whitten, E. H. T., 60
Williams, A., 27, 56
Wilson, G., iv
Wisser, E., 42
Wollnough, W. G., 7
Wulff net, 5, 6, 60, 69, 87

Zimmer, P. W., 31